The Beaver

The Beaver

Natural History of a Wetlands Engineer

Dietland Müller-Schwarze and Lixing Sun

Comstock
Publishing
Associates,
a division of Cornell University Press, Ithaca and London

This book has been published with the aid of a grant from the Humane Society of the United States.

Art for beaver tracks used courtesy of Kim A. Cabrera.

First published 2003 by Cornell University Press

Printed in the United States of America

Library of Congress Cataloging-in-Publication Data

Müller-Schwarze, Dietland.
The beaver : natural history of a wetlands engineer / Dietland Müller-Schwarze and Lixing Sun.
p. cm.
Includes bibliographical references (p.).
ISBN 0-8014-4098-X (cloth)
1. Beavers. I. Sun, Lixing. II. Title.

QL737.R632M85 2003
599.37--dc21 2002041522

Cornell University Press strives to use environmentally responsible suppliers and materials to the fullest extent possible in the publishing of its books. Such materials include vegetable-based, low-VOC inks and acid-free papers that are recycled, totally chlorine-free, or partly composed of nonwood fibers. For further information, visit our website at www.cornellpress.cornell.edu.

Cloth printing 10 9 8 7 6 5 4 3 2 1

Contents

Preface

With this book, we intend to fill a serious gap in the wildlife literature. During courses in the field and on campus, undergraduate students have asked for an up-to-date overview on the general behavior and ecology of the beaver. They could find none in the libraries. Nor have public libraries a modern, comprehensive book on beaver biology that would be helpful for high school projects. The scientific information on beavers is scattered over a vast universe of journals.

And yet, the beaver remains an extremely popular animal. In our state parks, guided nature tours to beaver ponds attract the most visitors, and the species ranks high in popularity among schoolchildren. Moreover, antitrapping sentiment on both sides of the Atlantic Ocean has led to sagging fur prices, less interest in fur trapping, and mushrooming beaver populations. These numerous beavers cause much wildlife damage, especially where suburban sprawl invades their habitat or where beavers recolonize their former, but now developed, haunts. In many places the beaver has become a political animal, pitting landowners against animal rights' advocates and wildlife managers against concerned citizens. To avoid costly and frustrating mistakes, wildlife managers, property owners, and other dedicated laypersons need an up-to-date book on beaver biology and management, including conflict resolution.

On both sides of the Atlantic, a number of symposia and workshops in recent years have been dedicated to the subject of beavers. Much of what we include in this book we have presented before in both written and spoken form at several such meetings. The contact with colleagues, wildlife managers, property owners, highway superintendents, and citizens' groups made us painfully aware of the problems and taught us to think about which parts of the biology and ecology of the beaver might be essential to know for wise management.

The book covers both species of beavers. Given our North American perspective, the primary focus is on *Castor canadensis*. The material encompasses recent research results—that is, most cited references date from the 1960s to the year 2001.

Many colleagues and helpers have contributed to the original data covered in this book. They include former graduate students Drs. Bruce Schulte and Peter Houlihan, Heather Brashear, Meaghan Boice-Green, Maryann Schwoyer, Rebecca Coleman Quail, Brett Mosier, Kristen Buechi, and Stasia Bembenek. Numerous undergraduate students and field assistants have helped over the years. Staff at Allegany State Park, Huntington Wildlife Forest, notably Tim Schwender and the late Dick Sage, and Cranberry Lake Biological Station, especially Larry Rathman, enabled our work in various ways. Uncounted callers with "beaver problems" have motivated us to feed the results of our academic studies into the mainstream of nature conservation and practical beaver management. Basic and applied research are not so separate after all.

Introduction

How many pelts in a beaver coat?
A beaver as large as a bear?
Beaver declared a fish?
What is "beaver fever"?
The beaver in a court of law?
Do otters prey on beavers?
Engineers lose jobs to beavers?
Beavers a menace to trout streams?

Find the answers in the pages ahead.

Two beaver species inhabit our world: the North American and the Eurasian beaver. Both had been extirpated over large areas by the beginning of the 20th century. But during the past 50 years, and continuing today, each of the species has traveled along a different trajectory. In the United States, reintroduction of the North American beaver in its former range has been so successful that burgeoning populations have no choice but to move into developed land. Such "nuisance beavers" make headlines by flooding land and downing trees. Like deer and Canada geese, the beaver has joined the ranks of overabundant wildlife.

In Europe, meanwhile, reintroductions have given some countries their first beavers in decades. Still small in numbers, these new populations are being carefully nurtured. Given that much of the landscape is developed, the "carrying capacity" will be reached there much sooner than in North America.

"Viewable wildlife," as an alternative to hunted and trapped game species, is gaining in importance with increasing urbanization and antitrapping and antifur sentiment. Beavers are singularly suited as viewable wildlife because they can be found at the same place year-round. First-hand experience with these animals leads to an understanding of the profound, but not easily visible, ecosystems services they provide. This represents a sea change in our attitude toward wildlife. Historically valued for their fur, meat, and perfume and medical ingredients, beavers are now appreciated alive more than dead.

Watching beavers for the first time or finding some moving into or through one's property raises many questions about their natural history. In our unpublished survey in middle and high schools in upstate New York, for example, children listed more harmful effects of beavers than benefits. Ecosystem services beavers provide, such as creating wetlands that purify water and increase biodiversity, need explaining and education. This is what this book is all about. The following chapters illuminate what beavers are doing and provide insight into what they could be up to next.

The text is organized into five parts. It progresses from the beaver as a single organism in Part I to the behavior of family units and the celebrated artifacts such as dams and lodges that result from this behavior in Part II. Part II also discusses communication by smell, sound, and visual signals, as well as the feeding behavior that is so obvious to even a casual observer by the sightings of felled trees and remaining stumps. Part III expands our view to whole populations, while Part IV explains specific aspects of the beaver's relationships to particular components of its habitat, such as water, vegetation, predators, and diseases. The beaver, in turn, creates habitat for other organisms, also covered in Part IV. Finally, Part V turns to the relationship between humans and beavers, to many readers perhaps the most timely topic. Taking the historical approach, this section of the book reviews fur trapping, the fur trade, and its historical impact: trapping depleted populations; reintroductions replenished the stocks. Part V also discusses the problem of "nuisance beavers," how to deal with them by proactive management, and how we might harness the beavers' "ecosystem services," to serve us as "ecosystems engineers."

Part I

The Organism

chapter 1 Now and Then: The Species, Including Fossils

In creation stories of many northern native Americans, at the time of beginning and transformation, giant animals roamed the world and ate people.

The culture hero Saya of the Athapaskan-speaking "Beaver Indians" at the Peace River in British Columbia and Alberta took on, and transformed the giant, people-eating animals to their current size. Now the roles are reversed: people eat the animals.

The Amikwa on the north shore of Lake Huron have the Great Spirit Manitou shrink the giant animals to their present size.

R. Ridington, 1981

Comment: The sight of large bones of Pleistocene mammals might have spawned such narratives to link the past to the present. Ethnographers have also often invoked a "collective memory" stemming from times when giant animals coexisted with people.

Authors

Living Beavers

The beaver is the second largest rodent after the South American capybara (*Hydrochoerus hydrochaeris*). Beavers belong to the family Castoridae in the suborder Sciuromorpha of the order Rodentia. They are more closely related to squirrels and marmots than to mouselike rodents (Muridae). For expediency, clever politicians have maneuvered the beaver into strange taxonomic neighborhoods. In 1760, the College of Physicians and the Faculty of Divinity in Paris permitted beaver meat on fasting days because the beaver's scaly tail classified it as a fish and not a mammal! By the way, the beaver is not alone in this regard. The pope declared the similarly semiaquatic capybara of South America, the largest rodent in the world, a fish, when petitioned by Venezuelans and Colombians in the 16th century. So every year observant parishioners eat 400 tons of capybara during the fast of Lent.[1]

Two species of beaver live today: the North American *Castor canadensis* and *Castor fiber* in Eurasia. Some taxonomists consider all beavers in the world to be one species. The North American beaver occurs from coast to coast and ranges from Alaska, Hudson Bay, and northern Labrador in the North to the U.S.-Mexico border, Gulf Coast, and Florida state line in the South (Fig. 1.1).

In the Old World, the beaver used to occur throughout Europe to the Mediterranean. Today it exists in small but expanding pockets in Scandinavia, Poland, and Russia, and along the Rhone and Elbe Rivers, and has been reintroduced in many areas, including along the Danube and its tributary, the Inn, and several other river systems in eastern Germany. A number of countries, such as Switzerland, have also reintroduced beavers, while in the Czech Republic beavers have immigrated voluntarily (V. Kostkan, personal communication, 2000) (Fig. 1.2).

The North American beaver cuts almost all tree species and builds elaborate freestanding lodges. The like-sized Eurasian beaver, on the other hand, cuts mostly willow, even in mixed forest stands, and lives in bank burrows. For instance, along the Danube near Vienna, Austria, researchers found only one freestanding lodge; all others were bank lodges (J. Sieber, personal communication, 1982). The Eurasian beaver appears more ancient and conservative than its New World cousin, thought to be a younger and more progressive species.

The two beaver species differ in their number of chromosomes: *C. canadensis* has 40 chromosomes (2n), and *C. fiber* has 48 (2n) (see Table 1.1).[2] The two species can also be readily distinguished by alleles of the esterase-D locus (Es-d): on horizontal starch gel electrophoresis, the Eurasian beaver exhibits a fast-migrating allele, while the North American beaver has a slowly migrating allele.[3]

Fossil Beavers

Enough fossils exist to illuminate the ancestry of modern beavers. The smallest extinct beavers were the size of a muskrat, and the largest may have reached the size of a black bear, although this is disputed now. During the Pleistocene, the age of giant mammals, huge beavers lived side by side with other colossal mammal species. It is anybody's guess what their lodges must have looked like, if they built any. Chances are they did not construct lodges. Table 1.2 summarizes the sequence of beaver fossils found in North America and Eurasia. Studies on fossils show that beavers occurred from the early Oligocene to the Holocene in the New World, the late Oligocene to the Holocene in Europe, and the late Miocene to the Holocene in Asia.[4]

The first clear castorid was *Palaeocastor* from the Upper Oligocene of North America. Although clear evolutionary lines cannot be established, some special-

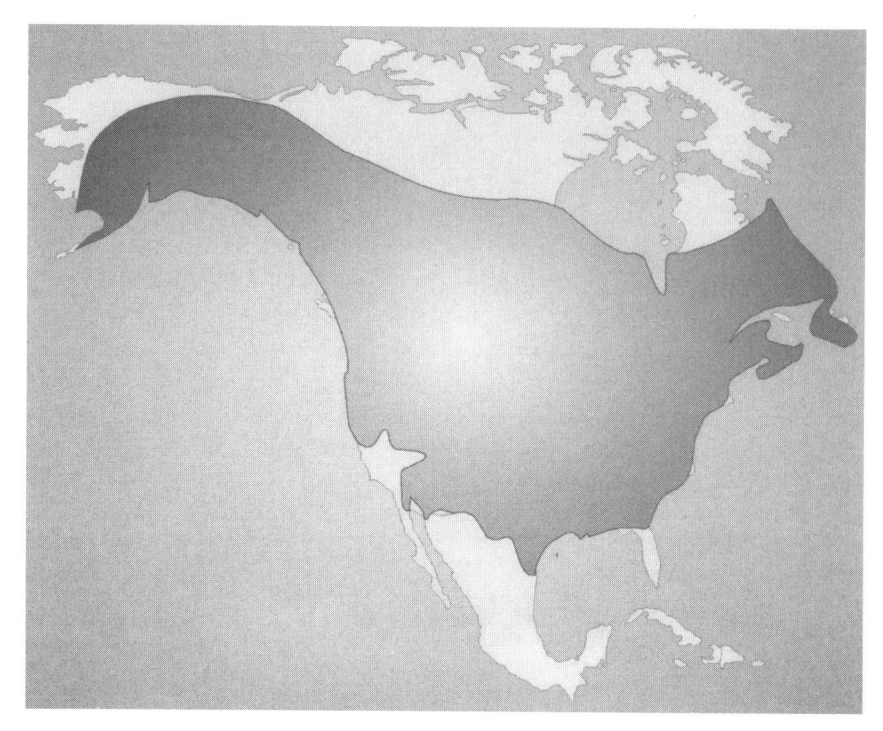

Figure 1.1 | Distribution of the North American beaver.

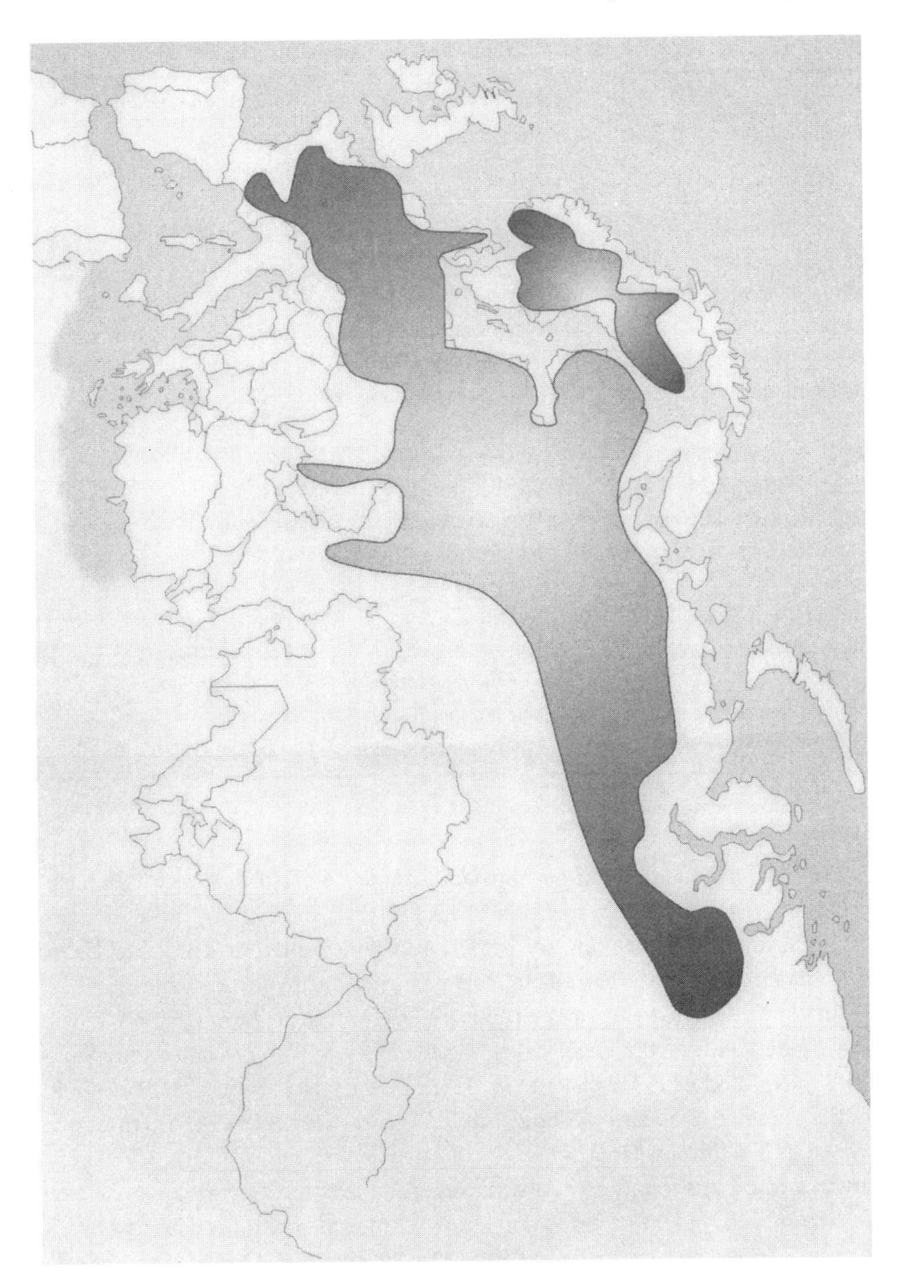

Figure 1.2 | Distribution of the Eurasian beaver.

Table 1.1 | Some Differences between the Two Beaver Species

	Eurasian Beaver	North American Beaver
Chromosomes	2n = 48	2n = 40
Body size		Older beavers can be slightly larger
Skull: nasal opening	Triangular	Square
Skull volume	Smaller	Larger
Uterus masculinus	Present	Absent
Tail dimensions	Narrower: width about 47% of length	Broader: width about 56% of length
Tail vertebrae	Narrower; processes less developed	Broader; with processes for tail muscle attachment
Anal gland secretion	Darker in females	Darker in males
Average litter size	Smaller: 1.9–3.1	3.2–4.7
Resistance to tularemia	Well developed	Very weak
Intestinal helminth *Travossosius rufus*	Not found in western Europe	Prevalent
Beaver beetle *Leptinillus validus*	Absent	Present
Dam building behavior	Less developed	More sophisticated
Lodges	Mostly bank lodges	Many freestanding lodges
Competitiveness	Less competitive	
Scent mounds	Smaller	Some "giant scent mounds"

Sources: References 4, 5, and 6.

ists speak of a sequence *Eucastor-Dipoides-Castoroides* in North America. In Eurasia, some have thought of a sequence *Steneofiber-Palaeomys-Castor.*

In the middle and late Tertiary, beavers were very numerous. The genus *Castor* may have migrated from Eurasia to North America during the Pliocene.

In North America, most beaver fossils stem from the eastern part, notably in the region south of the Great Lakes. *Castoroides* fossils were contemporaneous with the mastodon, as for instance in Wayne County, New York. *Castoroides* lived in the Pleistocene, as early as about 1 million years ago. *Castoroides,* up to 2.75 m long, was 5–6 times as large as our North American beaver today. In 1867 beaver-gnawed wood was found near Albany, New York, in the same cavity as a skeleton of a mastodon, and 5 feet (~150 cm) above it. Imbedded in clay and river ooze, it rested on gravel and was covered by peat. In addition to *Castoroides, Castor californiensis* was another large species.

In Europe, *Trogontherium* ("gnawing animal") fossils surfaced first in the Sea of Azof and later in England. Fossil beavers are commonly found in peat bogs of England and Ireland, together with the famous Irish elk, *Megaceros.* Remains of *Castor fiber* rested next to those of hippopotami, rhinoceros, and hyenas in a

Table 1.2 | Fossil Beavers

Years before Present (BP)	Earth Age	North America	Eurasia
	Pleistocene		
1.9 million–8,000		*Castoroides ohioensis*[a] *Castor canadensis*[b] spreads	*Trogontherium*[c] *Castor fiber*[b]
	Tertiary		
5.3–1.9 million	Pliocene	*Dipoides* *Amblycastor* *Castor* arrives in North America, possibly via North Pacific land bridge *Eucastor*	*Trogontherium* *Steneofiber* *Castor* sp.
23.9–5.3 million	Miocene	*Eucastor*	*Castor* sp. (late Miocene) *Palaeomys* *Steneofiber eseri*[d]
33.7–23.9 million	Oligocene	*Palaeocastor* *Agnotocastor*	*Steneofiber fossor*[e]
55–33.7 million	Eocene	North America and Eurasia separate Explosive evolution of rodents Earliest beaver-like fossils in California, Germany, and possibly China	
66–55 million	Paleocene		

Source: Reference 7.

[a] Largest known member of beaver family (Castoridae). Its skull was 34 cm long.

[b] Both species of Castor coexisted with the giant forms until about 10,000 years ago. *C. fiber* and *C. canadensis* have been isolated for at least 9000, possibly as long as 24,000 years.[3]

[c] Resembles more today's *C. fiber* than *Castoroides.* Others claim that the two giant forms are nearer to each other than to either of the two living *Castor* species.

[d] Not adapted to aquatic life.

[e] Led underground life.

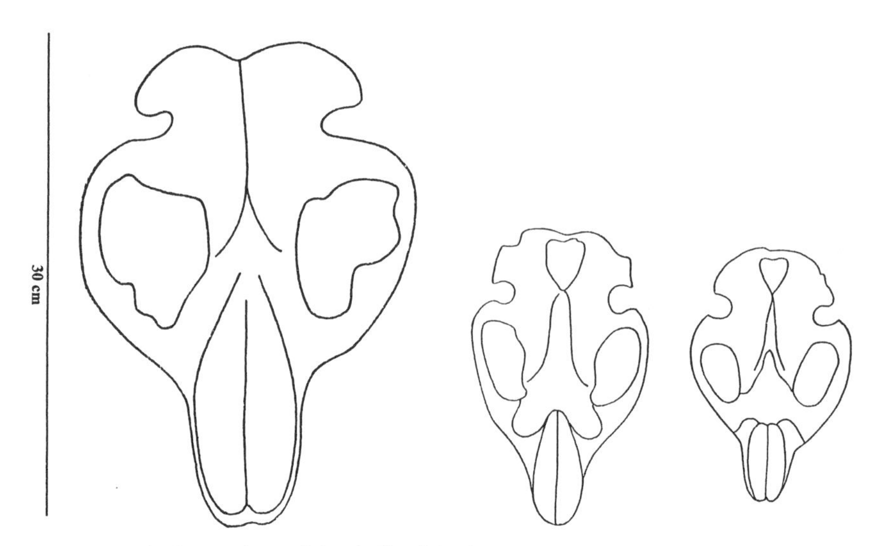

Figure 1.3 | Comparison of the skulls of the fossil giant beaver *Castoroides ohioensis* (left) and the extant Eurasian (*Castor fiber*) (middle) and North American (*Castor canadensis*) (right) beaver. Proportions (e.g. nasal cranium) are more diagnostic than any size differences. From: Hinze, G. 1953. Der Biber.

Pleistocene formation of Italy's Val d'Arno. Giant beaver fossils occurred close to those of the smaller *Castor* species as we know them today, as was the case with fossils of *Trogontherium* and *C. fiber* near Liège in Belgium.

It has been suggested that beavers of the Tertiary migrated between Eurasia and North America in both directions: *Steneofiber* from Eurasia to America, and *Eucastor* from the New World to the Old.[7] Several *Eucastor* species were extinct already during the middle Pliocene.

During the Pleistocene, there were also beavers greatly exceeding the size of present beavers, such as the now extinct *Trogontherium* in Europe. *Castoroides ohioensis* in North America was even larger. It lived about a million years ago.[8] Recent recalculations have refuted the traditional claim that *Castoroides* was as large as a black bear. That earlier conclusion had been reached by extrapolating from the size relationships of skulls and femurs to body size in living beavers to the fossil ones. Because the skull evolves in response to special needs, such as gnawing and chewing, it is a poor predictor for body mass. One obtains a different formula for the relationship between these measures and body mass by using the skull and femoral lengths of a broader taxonomic group, such as 19 rodent species.[9] According to new modeling, the body mass of *Castoroides* was probably 60–100 kg (*Palaeocastor*: 0.8–1.2 kg).[9] *Castor californiensis* was another large species. Figure 1.3 compares the skull sizes of *Castoroides*, the North American, and the Eurasian beaver.

REFERENCES

1. Mukerjee, M. 1994. What's in a name? When capybaras become fish and tomatoes are vegetables. Scientific American 271 (4): 26.
2. Ward, O. G., A. S. Graphodatsky, D. H. Wurster-Hill, V. R. Eremina, J. P. Park, and Q. Yu. 1991. Cytogenetics of beavers: a case of speciation by monobrachial centric fusions. Genome 34: 324–328.
3. Sieber, J., F. Suchentrunk, and G. B. Hartl. 1999. A biochemical-genetic discrimination method for the two beaver species, *Castor fiber* and *Castor canadensis*, as a tool for conservation. In: P. E. Busher and R. M. Dzięciołowski, editors. Beaver protection, management, and utilization in Europe and North America. New York: Kluwer Academic/Plenum. p 61–65.
4. Heidecke, D. 1986. Taxonomische Aspekte des Artenschutzes am Beispiel der Biber Eurasiens. Hercynia 22: 146–161.
5. Pilleri, G., C. Kraus, and M. Gihr. 1985. Ontogenesis of *Castor canadensis* in comparison with *Castor fiber* and other rodents. In: Investigations on beavers. Volume IV. Berne: Brain Anatomy Institute. p 11–82.
6. Rosell, F. and K. V. Pedersen. 1999. Bever. Aurskog, Norway: Landbruksforlaget.
7. Pilleri, G. 1983. Introduction: phylogeny, systematics, geographical distribution. In: G. Pilleri, editor. Investigations on beavers. Volume I. Berne: Brain Anatomy Institute. p 9–15.
8. Taylor, W. P. 1916. The status of beavers in western North America, with a consideration of the factors in their speciation. University of California Publications in Zoology. 12: 413–495.
9. Reynolds, P. S. 2002. How big is a giant? The importance of method in estimating body size of extinct mammals. Journal of Mammalogy 83: 321–332.

chapter 2

Form, Weight, and Special Adaptations

> . . . their powerful incisor teeth not only serving them to strip off and divide the bark of trees, but also enabling them, when urged by their instinct of construction, to gnaw through trunks of considerable thickness, and thus obtain the timber of which they stand in need for the building of their habitations. These important organs contribute, therefore, in an especial manner, to supply them both with food and shelter.
>
> *E. T. Bennett, 1835*

Size and Body Weight

The weight of an adult North American beaver ranges between 40 and 50 lb (~18–23 kg). Recorded beaver weights are 96 lb (43.6 kg) in 1960 in Missouri[1] and about 110 lb (50 kg) in 1921 at the Iron River in Wisconsin.[2] The body including the tail reaches about 48 inches (1.2 m) in length. The tail itself is about 16–17 inches (~40 cm) long, about 6–7 inches (16 cm) wide, and ¾ inch (1.9 cm) thick.

Tail

Most distinctive for the beaver, the tail is flat and scaly (See Fig. 3.2). It is a multipurpose tool. Beavers use their tail as prop to balance themselves when cutting trees. It is also important as a rudder during diving and underwater maneuvering around all three body axes. The tail signals alarm when it is slapped on the water surface (see chapter 6). In addition, fat reserves are stored in the tail. It also serves in heat exchange through a countercurrent arrangement of blood vessels[3] (see chapter 3). By this mechanism the beaver can reduce the 25% heat loss via the tail in the summer to 2% in the winter.[4]

Nostrils, Ears, and Eyes

When the beaver is under water, valves close the nostrils and ears. Fur-lined lips can be closed behind the teeth. This permits gnawing under water. Part of the tongue and the epiglottis prevent water from entering the larynx and trachea. "Diving goggles," a nictitating membrane, protects the eyes under water.

Brain

The brain of a 17-kg North American beaver weighs about 45 g, and that of an 11.7-kg animal, 41 g.[5] The size of the brain relative to body weight is considered to indicate how well a species processes information and solves complex problems posed by the environment, often shortened to the term *intelligence.* To determine how "intelligent" the beaver is in relation to other related rodents of similar size, the brains are compared using an *encephalization quotient* (EQ). This quotient relates the measured brain size (weight) of a species to the expected brain size. The expected brain size is an average for all rodents (or all mammals), computed as brain weight in relation to body weight. The EQ of the aquatic beaver (0.9) is intermediate between that of terrestrial rodents of similar size and arboreal rodents such as squirrels. Muskrat and nutria have lower EQs (0.679 and 0.779, respectively). The EQ (0.8) of the Eurasian beaver is slightly lower than that of the North American beaver.[5]

The size and anatomical appearance of the various brain parts do not betray any specific adaptations to the beaver's semiaquatic lifestyle. For instance, the olfactory bulb of beavers is neither smaller nor larger than that of other mammals of similar size that are not adapted to water. The ratio of the length of the olfactory bulb to the length of the cerebrum is 0.35, the same as in the Alpine marmot (*Marmota marmota*) and very similar to the ratio in the 13-lined ground squirrel (*Citellus tridecemlineatus;* 0.33) and the common European red squirrel (*Sciurus vulgaris;* 0.34).[6]

Certain brain measurements have been used to infer intelligence. If the evolutionarily older hypothalamus is relatively small in comparison to the cerebrum, which contains the "neocortical" higher centers, the brain is considered more advanced. Among a number of related squirrel-like rodents, the beaver scores highest in this regard. The ratio of hypothalamus length to cerebrum length ranged from 0.20 to 0.24 in 12 North American beavers from Mississippi.[6]

The cerebellum, involved in the coordination of locomotion in three-dimensional space, is well developed, although arboreal rodents such as squirrels have equally large, if not larger cerebellum hemispheres.[6]

Large areas of the brain (neocortex) are dedicated to processing somatic sensory and auditory stimuli. Within the main somatic sensory cortex, the lips and hands are more strongly represented than the tail and vibrissae.[7] In the less aquatic capybara, these areas are smaller. The visual area, by contrast, is small in the beaver but most developed in arboreal rodents such as the gray squirrel (*S. carolinensis*).[5]

Skull and Teeth

The skull is massive; it has to withstand the forces of powerful chewing (masseter) muscles. The four incisors (Plates 1 and 2) are formidable chisels that grow

Figure 2.1 | The toenail of the second toe on the hind foot (on right) is split, thought to be important as a "comb" for preening the fur.

continuously. Their outer enamel layer appears bright orange. The roots of the lower incisors take up most of the length of the lower mandible. Adapted to gnawing wood, the tooth is very hard, ensured by a high ratio of outer enamel thickness to total enamel thickness.[8] Each side of the upper and lower jaw carries 1 premolar and 3 molars, for a total of 20 teeth. Meandering enamel ridges on the chewing surface of the molars act as grating instruments for shredding woody food.

Feet

The handlike front feet are dexterous. They are used to grasp and manipulate food, dig, and groom the fur. The hind feet are webbed and much larger. In the water, they propel the body. The second toe of each hind foot has a split nail. It serves as a "comb" for preening the fur to keep it fluffy (Fig. 2.1). Beavers leave a characteristic track, especially in soft snow (Fig. 2.2)

Fur

The fur consists of two types of hair—the long, coarse guard hairs and the soft wool hair of the underfur, right close to the skin. The guard hairs are 2.0–2.4 inches (5–6 cm) long, and the underfur is 0.8–1.2 inches (2–3 cm) long. It is the layer of wool hair that keeps the body warm. As Hilfiker[9] pointed out, the beaver's fur serves as a warm overcoat, as a raincoat when under water, as a lifejacket to

Figure 2.2 | Beaver tracks in snow.

keep its bearer afloat, and as protection against the teeth and claws of enemies. The fur is extremely dense: with 12,000–23,000 hairs/cm^2, the beaver has more hair per area of skin than the nutria from warmer zones (8,000–13,000) but less than the river otter (25,000–51,000). Beavers molt their hair once a year during summer. It is in "prime condition" between December and March, when pelts are most desirable to fur trappers.

Digestive Tract

Beavers have a digestive tract adapted to herbivory. The lesser curvature of the stomach has a peculiar gland patch, the cardiac gland, that secretes into the stomach. Since it is also found in the koala (*Phascolarctus*) and wombat (*Vombatus ursinus*), this gland may have to do with a high-fiber diet. The intestine is very long, about 6 times the body length. The caecum is very large, its volume twice the size of the stomach, and its length five-sevenths of the body length.

Excretion

The beaver produces dilute urine. Concentrated in the "castor sacs," it becomes castoreum, used for scent marking (see chapter 6). The feces appear as compacted balls of sawdust and are deposited in water (Plate 3). Both pass through the "cloaca," a single opening for digestive, excretory, and genital tracts, as well as the scent-producing castor sacs and anal glands.

Reproductive Organs

Both sexes look alike externally. We can tell the sex of a beaver by close examination of an immobilized animal: with the beaver lying on its back, palpating the abdomen close to the root of the tail reveals the cartilaginous *baculum,* also known as "penis bone," or *os penis,* in a male. All sex organs are internal. The male North American beaver has darker and more viscous anal gland secretions than the female,[10] while the reverse is true for the Eurasian beaver, where males have a vestigial "uterus masculinus."[11] The testes are relatively small, comprising 0.05% of the body weight.[12] This compares with 0.08% in humans. The "relative testes size" is the ratio of observed mass to that predicted by the general mammalian formula. For mammals in general, $Y = 0.035X^{0.72}$, where Y is the weight of both testes and X the body mass. For rodents: $Y = 0.031\ X^{0.77}$.[13] According to these formulae, a 25-kg beaver should have a relative testes size of 0.36 or 0.37, but its actual value is 0.22.

Females have four mammae, located in the chest region. During lactation, they are easily visible and can be used for sex identification in the field.

Milk

The beaver's milk is very rich. It contains 19% fat. This is high compared with other rodents, higher than in lagomorphs, close to the percentage in the brown bear, and clearly surpassed only by that in pinnipeds. The 11.2% protein content is also high by rodent standards, even higher than the protein content in most pinnipeds and second only to the eastern cottontail rabbit. Sugar content (1.7%), on the other hand, is low. It is lower only in pinnipeds. The ash content in beaver milk (1.1%) is average for mammals, and its overall energy content (92.46 kcal/gram) is high, among wild mammals surpassed only by that for pinnipeds.[14,15]

The Senses

The sense organs are aligned in a row so that beavers swim with their nostrils, eyes, and ears raised above the waterline while the rest of the head and body is submerged (See Fig. 3.1). This arrangement resembles that in other semiaquatic animals such as hippopotami or alligators.

Smell

The sense of smell is acute, as witnessed by how the beaver sniffs toward any disturbance, such as humans near a beaver lodge. Beavers live a crepuscular and nocturnal life, so olfaction is the sense of choice. They smell their food before they choose it. Field experiments conducted with paper bags over tree seedlings demonstrated that beavers do not need to see their potential food to select the palatable kind.[16]

The nose is also extremely important in social communication. Beavers sniff each other, and particularly the scent marks located on mud mounds, to extract information. Beavers communicate extensively with the aid of castoreum and anal gland secretions, illustrating how beavers rely on chemical cues in their social life. Chapter 6 on scent communication details the great variety of chemical signals that beavers broadcast via their scent mounds.

Beavers also sniff the air for predators and any other strange objects and odor sources that may indicate danger.

Vision

The eyes are small, and accordingly, the beaver's eyesight is considered poor. In their orientation, they conform to those of other "pursued" mammals (i.e., prey animals). To provide a large field of vision, the angle between the optical axis of the eye and the body axis is as large as 65 degrees. This resembles the 85 degrees in rabbits. Extreme in this regard is the horse. Its angle of monocular vision is 215 degrees! In predators, on the other hand, the eyes diverge only a little, in cats only 4–9 degrees. Finally, primates' eyes have parallel axes, which allow for the three-dimensional vision they need to manipulate objects or to swiftly move through trees.

Is the beaver's eye especially adapted to seeing under water? The cornea of the eye refracts light when the animal is in air, but under water it becomes a clear window lacking refractive properties. To compensate for this, aquatic birds and mammals that hunt underwater, such as cormorants, mergansers, dippers, and river otters, change the curvature of their lenses when submerged. They accomplish this with powerful muscles attached to the lens—the ciliary and iris muscles. Compared to the river otter, however, the sphincter and dilator muscles in the eye of the beaver are rudimentary. Although the beaver operates underwater, as when feeding on the submerged food pile, its eye is not particularly adapted to underwater vision, as it is in semiaquatic carnivores.[17]

Hearing

Since beavers use auditory signals such as the tail slap, and whining and hissing, their hearing is functional, although we don't know the precise sensitivity and performance of their acoustical sense. We also have to consider water as a medium for sound propagation, in addition to air.

REFERENCES

1. Rue, L. L., III. 1968. World of the beaver. Philadelphia: J. B. Lippincott.
2. Bailey, V. 1927. Beaver habits and experiments in beaver culture. U.S. Department of Agriculture Technical Bulletin 21. Washington, D.C.: U.S. Department of Agriculture.
3. Cutright, W., and T. McKean. 1979. Countercurrent blood vessel arrangement in beaver (*Castor canadensis*). Journal of Morphology 161: 169–176.

4. Marchand, P. J. 1996. Life in the cold. Lebanon, New Hampshire: University Press of New England.
5. Pilleri, G., M. Gihr, and C. Kraus. 1984. Cephalization in rodents with particular reference to the Canadian beaver (*Castor canadensis*). In: G. Pilleri, editor. Investigations on beavers. Volume II. Berne: Brain Anatomy Institute. p 11–100.
6. Pilleri, G. 1983. Central nervous system, cranio-cerebral topography and cerebral hierarchy of the Canadian beaver (*Castor canadensis*). In: G. Pilleri, editor. Investigations on beavers. Volume I. Berne: Brain Anatomy Institute. p 19–59.
7. Carlson, M., and W. I. Welker. 1976. Some morphological, physiological and behavioral specializations in North American beavers (*Castor canadensis*). Brain, Behavior and Evolution 13: 302–326.
8. Korvenkontio, V. A. 1934. Mikroskopische Untersuchungen an Nagetierincisiven, unter Hinweis auf die Schmelzstruktur der Backenzähne. Annales Societatis Botanicae Fennicae 'Vanamo' 2: 1–274.
9. Hilfiker, E. L. 1990. Beavers: water, wildlife and history. Interlaken, N.Y.: Windswept Press.
10. Schulte, B. A., D. Müller-Schwarze, and L. Sun. 1995. Using anal gland secretion to determine sex in beaver. Journal of Wildlife Management 59: 614–618.
11. Rosell, F., and L. Sun. 1999. Use of anal gland secretion to distinguish the two beaver species *Castor canadensis* and *C. fiber*. Wildlife Biology 5: 119–123.
12. Osborn, D. J. 1953. Age classes, reproduction, and sex ratios of Wyoming beaver. Journal of Mammalogy 34: 27–44.
13. Kenagy, G. C., and S. C. Trombulak. 1986. Size and function of mammalian testes in relation to body size. Journal of Mammalogy 67: 1–22.
14. Zurowski, W., J. Kisa, A. Kruk, and A. Roskosz. 1974. Lactation and chemical composition of milk of the European beaver (*Castor fiber*). Journal of Mammalogy 55: 847–850.
15. Oftedal, O. T. 1981. Milk, protein and energy intakes of suckling mammalian young: a comparative study [Ph.D. dissertation]. Ithaca, N.Y.: Cornell University.
16. Doucet, C. M., R. A. Walton, and J. M. Fryxell. 1994. Perceptual cues used by beavers foraging on woody plants. Animal Behaviour 47: 1482–1484.
17. Ballard, K. A., J. K. Sivak, and H. C. Howland. 1989. Intraocular muscles of the Canadian river otter and Canadian beaver and their optical function. Canadian Journal of Zoology 67: 469–474.

Diving and Thermoregulation: From Land Mammal to Semiaquatic Design and Function

> They cannot dive long time under water but must put up their heads for breath, which being espied by them that beset them, they kill them with gun-shot, or pierce them with Otters speares.
>
> *Edward Topsell, 1607*

Morphological Adaptations

The beaver's body and how it functions can be understood as a compromise of life on land and life in the water. The basic mammalian design has evolved into a superb amphibious, semiaquatic animal. Most peculiarly, the tail is dorsoventrally flattened. Although helpful in diving, aquatic life does not mandate this; the muskrat, with a laterally compressed tail, occupies the same habitat and even forages inside beaver lodges.

The body of the beaver is drop-shaped. Other aquatic animals, such as fish, penguins, seals, or whales, share this result of convergent evolution. In addition to having valvular ears and nose, and transparent eye membranes (see chapter 2), the beaver has lips that can close behind the incisors. This permits it to gnaw underwater. The cloaca is the single opening for the reproductive tract, excretion, and scent glands. This arrangement is thought to minimize infections when swimming in foul waters.

The anatomy of the beaver betrays the habit of extensive floating on the water surface: as mentioned earlier, the vital air intake and sense organs are arranged so that the nostrils, eyes, and ears extend above the water while the rest of the head and body are submerged (Fig. 3.1).

Webbed hind feet serve in the water, while beavers use the nonwebbed forepaws for digging, grooming, grasping, and manipulating. Such compromises are typical for the amphibious but not fully aquatic beaver. Living in shallow water, surrounded by terrestrial habitat, the beaver has acquired swimming and diving prowess but still has maintained its ability to walk, even run, dig, and forage on land.

Diving

Beavers can stay underwater for usually up to 5–6 minutes, with 15 minutes being their limit.[1] Most ponds are shallow and beavers never have to dive very

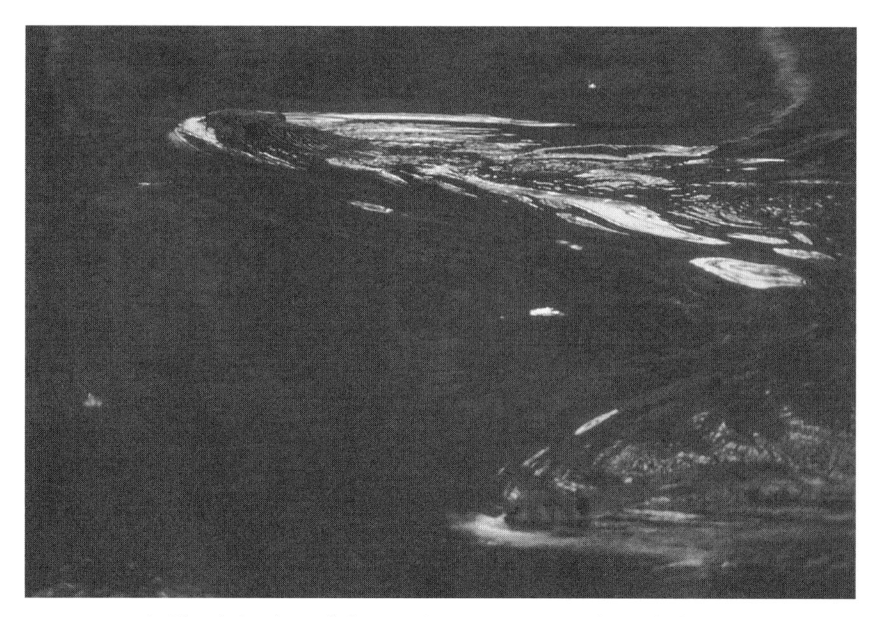

Figure 3.1 | The air intake and three major sense organs inform the beaver when swimming on the surface, with most of the body submerged.

deep. Diving forces the beaver to husband limited oxygen and requires special adaptations for thermoregulation under water. During a dive, the heart rate slows down (bradycardia), which saves oxygen. The bradycardia is accompanied by vasoconstriction. This "diving reflex" is triggered by nerve endings in the mouth and nose areas. As soon as the beaver's nose hits the water, blood rushes to the brain and heart, vital organs that would be damaged by an oxygen shortage. In addition to this increased flow of oxygen-rich blood to the brain, beavers tolerate high concentrations of carbon dioxide in their tissues.[3] The beaver's oxygen storage capacity is not exceptional. It resembles that of humans and is far less than that in harbor seals.[2] Its heart is relatively small, the hallmark of a slowly moving animal that does not exert itself for any length of time.[2] But beavers can exchange as much as 75% of the air in their lungs. By comparison, humans can exchange only about 15% per inhalation-exhalation cycle.

In a classic experiment, an 18-kg beaver was live-trapped in Canada's Algonquin Park. Resting in a box, its heart rate was 100 beats/minute. When it was set into water, the heart rate dropped to 50 beats/minute. Apnea, caused by pressure on the windpipe, also slowed the heart down. The beaver did not struggle when its head was immersed in water, but responded with a lower heart rate. After this simulated "dive," the heart rate rose again to 75–90 beats/minute and remained there for several minutes. This accelerated circulation helps to repay the "oxygen debt" from the dive.[1] A change from swimming to diving was accompanied by a change in heart rate from 125 to 67 beats/minute.[4]

The diving reflex found in beavers and other aquatic animals such as seals and muskrats has been put to good use in medicine. Humans also have immersion-sensitive nerve endings around the mouth and nose. Immersion of the face into cold water for 30–40 seconds slows down the heart rate by 10%–40%. In 90% of the tested patients the heart rate dropped after about 35 seconds. The colder the water, the greater the effect. This reflex lends itself to therapy. Abnormally fast heart rates of patients can be slowed down to a more normal rhythm by immersing their face in cold water, or even by placing a cold wet towel, rag, or plastic bag with ice water on their face. Children and young adults respond better than older adults. Compared with earlier methods, the procedure is noninvasive, simple, cheap, and repeatable over and over again. The method is useful as first aid but has to be followed up by long-term therapy.[5]

Thermoregulation in Water

Water is 800 times denser than air. Therefore, the fur of a diving mammal will be compressed in water, and the insulating air between the hairs will escape. The faster the beaver swims, and the deeper it dives, the more air will leak from its fur.[6] At the freezing point (0°C), the beaver's resistance to heat loss in water is about one-eighth of that in air at the same temperature. Beavers lose heat much more rapidly than do seals, whose resistance factor is only one-third. The polar bear resembles the beaver, rather than the seal, in this regard. Bear and beaver are land animals and not well adapted to prolonged stays in cold water.[7] The beaver's pelt accounts for only about 24% of its total insulation in water of 1°–4°C, body fat being responsible for the rest. The increased blood supply to the working muscles of the swimming beaver expands the thermal core of the body, leading to higher convective heat loss to the surrounding water.[6]

Tail and hind legs serve as heat exchangers (Fig. 3.2). The base of the tail contains intertwined networks of blood vessels.[8] In such a rete mirabile ("wonderful net"), arteries with warm blood from the core of the body transfer heat to closely located veins. The veins take the heat back to the core. This way, the blood at the surface of the tail is cooler, and little heat is lost to the surrounding cold water. In summer, a beaver can lose 25% of its body heat via its tail but only 2% in winter. In one experiment, the middorsal tail surface of a submerged beaver had the same temperature as the water, while the temperature under the dorsal skin was well regulated and only 4°–5°C below the abdominal temperature (36.7°C in water).[3] The fact that beavers can lose up to 20% of their generated heat via the tail[9] is important in climates with hot summers, as for example in Mongolia.[10]

In general, beavers cool off when foraging in the water, and raise their body temperature again when resting in the lodge. After the beaver forages under the

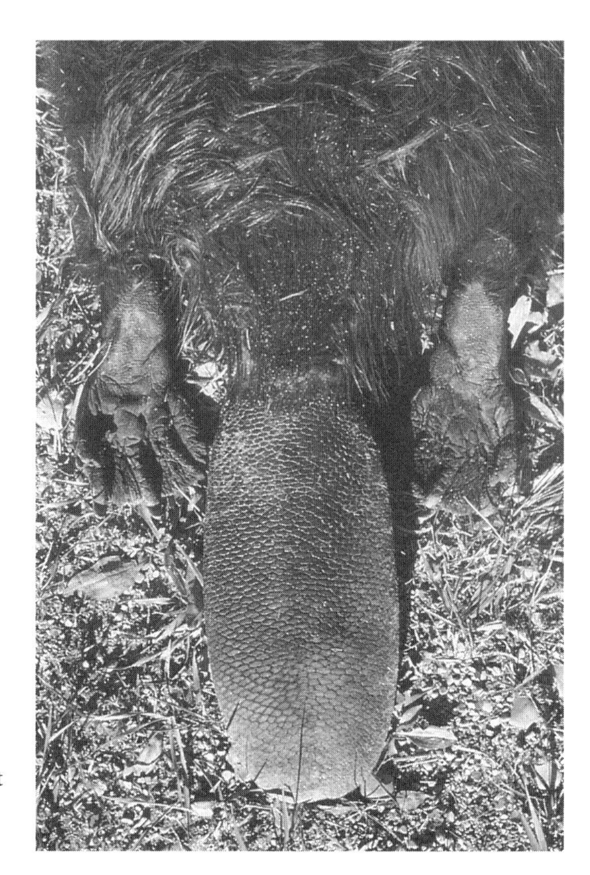

Figure 3.2 | The scaly tail of a beaver. Tail and hind feet regulate heat exchange with the environment.

ice for minutes, body temperature can drop below +34°C.[3] Kits cool faster than adults: in one study, the body temperature of adults dropped maximally 2°C after being immersed for 40 minutes, but kits lowered theirs by as much as 7°C after only 20 minutes![6]

The body temperature of free-ranging "transmittered" beavers dropped while they swam in the water after a stay in the lodge. During periods of immersion lasting 2–23 minutes, on average, their body temperature declined by up to 0.9°C. In the laboratory, adult beavers lowered their body temperature by up to 2.5°C over 40 minutes in water with a temperature of 2–3°C. In colder water the body temperature decreased more rapidly.[6] The cooling rates in the laboratory and in the field were the same for the same water temperature. After the beaver left the water, the body temperature recovered over a 60-minute period.

Core cooling in cold water can be retarded but not eliminated. Therefore, beavers must leave the water periodically. This amounts to behavioral thermoregulation. Accordingly, the younger kits (4–7 months old) spend shorter time periods (5–10 minutes) in the water than do adult beavers (14–23 minutes).[6]

Thermoregulation on Land

For comparison, on land the body temperature of an adult beaver oscillates around +37.1°C–37.3°C. It rises by 0.5°C when the beaver is active during the night hours from 1700 to 0200, but drops again below average at the end of the activity period, around 0200 to 0700.[11] Over a 24-hour period, adults varied their temperature by 1.4°C on average in summer and 2.5°C in winter.[12]

The body temperature of beavers varies seasonally. In autumn (late October to early November) the automatically recorded body temperature of three free-ranging adult beavers in Minnesota and Michigan averaged 36.3°C. Individual temperatures ranged from 34.5° to 37.6°C. The temperature declined from fall to winter at an average rate of 0.01°C/day. From November to April, the average body temperature was only 35.3°C (range: 32.5°–38.8°C). From early March on, the adults raised their body temperature at a rate of 0.03°C/day. A beaver kit, on the other hand, did not lower its body temperature over the winter. This appears to be related to the continuing body growth of kits during the winter. Body temperature in a yearling was intermediate between that in adults and the kit.[12]

REFERENCES

1. Irving, L., and M. D. Orr. 1935. The diving habits of the beaver. Science 82: 569.
2. McKean, T., and C. Carlton. 1977. Oxygen storage in beavers. Journal of Applied Physiology 42: 545–547.
3. Irving, L. 1937. The respiration of beaver. Journal of Cellular and Comparative Physiology 9: 437–451.
4. Gilbert, F. F., and N. Gofton. 1982. Heart rate values for beaver, mink and muskrat. Comparative Biochemistry and Physiology 73A: 249–252.
5. Weiner, C. 1988. Erste Hilfe—dem Biber abgeschaut. Münchener medizinische Wochenschrift 130(22): 20–22.
6. MacArthur, R. A., and A. P. Dyck. 1990. Aquatic thermoregulation of captive and free-ranging beavers (*Castor canadensis*). Canadian Journal of Zoology 68: 2409–2416.
7. Scholander, P. F., V. Walters, R. Hock, and L. Irving. 1950. Body insulation of some arctic and tropical mammals and birds. Biological Bulletin 99: 225–236.
8. Cutright, W. J., and T. McKean. 1979. Countercurrent blood vessel arrangement in beaver (*Castor canadensis*). Journal of Morphology 161: 169–176.
9. Steen, I., and J. B. Steen. 1965. Thermoregulatory importance of the beaver's tail. Comparative Biochemistry and Physiology 15: 267–270.
10. Stubbe, M., and N. Dawaa. 1986. Die autochthone zentralasiatische Biberpopulation.

Zoologische Abhandlungen, Staatliches Museum für Tierkunde, Dresden 41(7): 93–103.

11. MacArthur, R. A. 1989. Energy metabolism and thermoregulation of beaver (*Castor canadensis*). Canadian Journal of Zoology 67: 651–657.
12. Smith, D. W., R. O. Peterson, T. D. Drummer, and D. S. Sheputis. 1991. Over-winter activity and body temperature patterns in northern beavers. Canadian Journal of Zoology 69: 2178–2182.

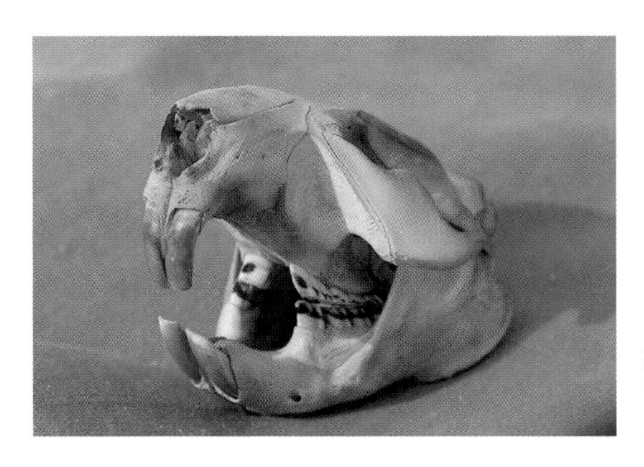

Plate 1 | The skull of a beaver, showing the large incisors. (See chapter 2.)

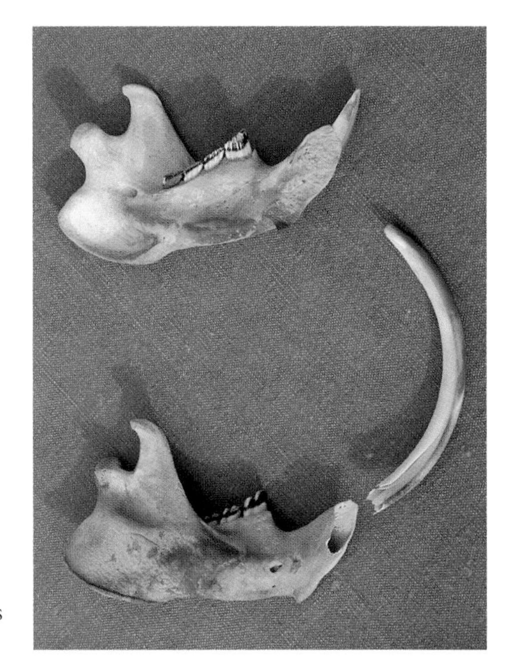

Plate 2 | The mandible accommodates most of the lower incisors. Top: incisor in normal position. Bottom: incisor pulled out of mandible to show its length. (See chapter 2.)

Plate 3 | Beaver droppings in water. (See chapter 2.)

Plate 4 | Food cache. Under the ice, beavers subsist on bark from this pile of stored branches near their lodge. (See chapter 4.)

Plate 5 | Three lodges in one pond. (See chapter 5.)

Plate 6 | Two beavers on a trail leading from the core area to their feeding grounds. (See chapter 5.)

Plate 7 | A parent hauls fresh foliage to the lodge as food for the kits. (See chapter 5.)

Plate 8 | Adult beaver and kit feeding in shallow water side by side. (See chapter 5.)

Plate 9 | A giant scent mound. (See chapter 6.)

Plate 10 | A beaver sniffs an experimental scent mound. (See chapter 6.)

Plate 11 | Sniffing is followed by pawing and marking the scent mound. (See chapter 6.)

Plate 12 | Castor sacs (top) and anal glands, or "oil sacs" (below) in a male. (See chapter 6.)

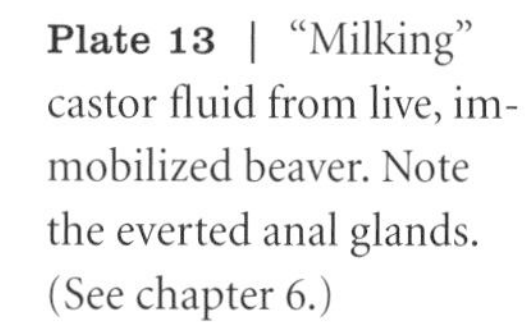

Plate 13 | "Milking" castor fluid from live, immobilized beaver. Note the everted anal glands. (See chapter 6.)

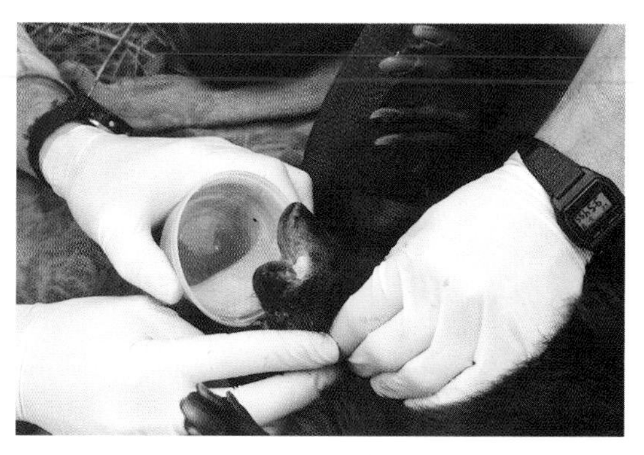

Plate 14 | Starting a dam with poles as braces against the stream bank. (See chapter 7.)

Plate 15 | Stones weigh down the logs and serve as permanent building blocks. (See chapter 7.)

Plate 16 | The crest of the dam is sealed with a thorough layer of mud. (See chapter 7.)

Plate 17 | A high dam at a steep stream. (See chapter 7.)

Plate 18 | Bank lodge. Bavaria. (See chapter 7.)

Plate 19 | Lake lodge without mud. Cranberry Lake, Adirondack Mountains, New York. (See chapter 7.)

Plate 20 | A thick layer of snow on a frozen pond, and the snow-covered frozen mud wall of their lodge, cut beavers off from the external light and temperature regimes. (See chapter 8.)

Plate 21 | Beavers hold sticks in their front paws and chew off the bark. (See chapter 9.)

Plate 22 | Scots pine felled and dismembered by beavers. (See chapter 9.)

Plate 23 | Red maple (in fall foliage) felled but not consumed by beavers. (See chapter 9.)

Plate 24 | Beaver sniffs and chooses experimental sticks laid out on the ground. (See chapter 9.)

Plate 25 | Beaver carries away a bundle of chosen sticks. The beaver will peel such short sticks in the safety of its lodge. (See chapter 9.)

Plate 26 | Array of experimental sticks after beavers have made their choices overnight. (See chapter 9.)

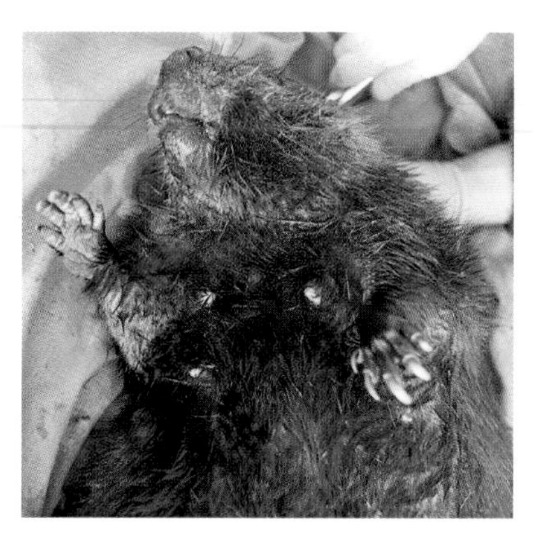

Plate 27 | Lactating females are easily identified by their enlarged nipples. (See chapter 10.)

Plate 28 | Beaver grooming itself. (See chapter 10.)

Plate 29 | "Starter home": young beaver, just dispersed, in its "token lodge" at a marginal site. This site was abandoned shortly thereafter. (See chapter 12.)

Plate 30 | Lake Tear of the Clouds, Adirondack Mountains, New York. Mt. Marcy, New York's highest mountain, in background. (See chapter 13.)

Plate 31 | Beaver site at Denali National Park, Alaska. (See chapter 13.)

Plate 32 | Beaver lodge in willow stand at Rocky Mountain National Park, Colorado. (See chapter 13.)

Plate 33 | Beaver dam in Zuni Mountains, New Mexico. (See chapter 13.)

Plate 34 | Developed landscape: bank lodge at backwater of River Elbe, Germany. (See chapter 13.)

Plate 35 | White suckers milling about below a beaver dam. Allegany State Park. (See chapter 16.)

Plate 36 | Canada goose near beaver lodge. (See chapter 16.)

Plate 37 | Mink track at beaver lodge. (See chapter 16.)

Plate 38 | Beaver meadow. (See chapter 16.)

Plate 39 | Field visits to beaver dams are popular with the public. (See chapter 16.)

Plate 40 | The coat of arms of the Hudson's Bay Company. *Pro pelle cutem* ("a skin for a skin") was its motto. (See chapter 17.)

Plate 41 | Tiles displaying beaver at the Astor Place Station of New York City's subway. (See chapter 17.)

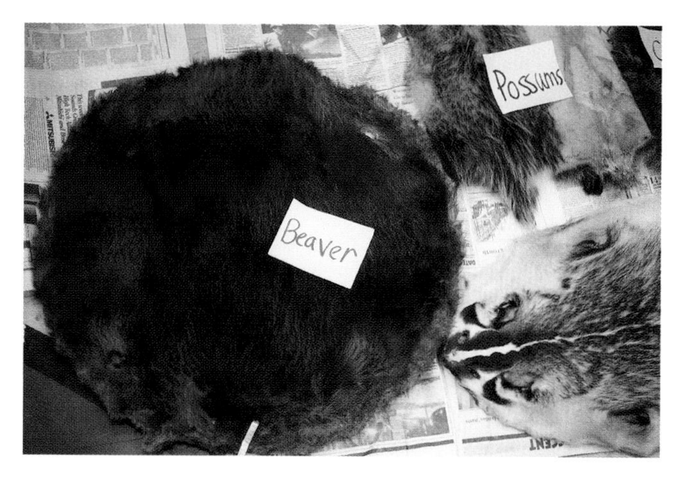

Plate 42 | Beaver skins are stretched over a circular frame for drying. (See chapter 17.)

Plate 43 | Historical marker in Old Forge, New York, commemorating the release of beavers into the Adirondacks in 1905. (See chapter 18.)

Plate 44 | Beaver lodge in southern beech (*Nothofagus*) forest in Tierra del Fuego, Argentina. (See chapter 18.)

Plate 45 | Road flooded by beaver dam in a ditch. (See chapter 19.)

Plate 46 | A degraded stream in Idaho, before introduction of beavers. Photo: Lew Pence. (See chapter 20.)

Plate 47 | Vegetation regrowth 3 years after introduction of beavers. Photo: Lew Pence. (See chapter 20.)

Plate 48 | Hancock live traps are used to trap and transfer beavers. Here two beavers were caught in one trap. (See chapter 21.)

Plate 49 | Traditional protection of specimen trees with welded wire is still very useful. Note the unprotected tree cut down by beavers. (See chapter 21.)

Plate 50 | Beaver Deceiver. (See chapter 21.)

Energy Budget

4 chapter

Beavers become very fat at the approach of autumn; but during winter they fall off in flesh, so that they are generally quite poor by spring.

J. J. Audubon and J. Bachman, 1854

Clearly, the beaver appears to be uniquely situated as a metabolic curiosity, particularly during the winter.

N. S. Novakowski, 1967

Energy Content of Food

Beavers work very hard. They modify their habitat extensively. How much energy does their proverbial and incessant activity require, and how do they meet these requirements? Some of the earlier measurements and estimates of food consumption by beaver were done between 1938 and 1962.[1–6]

Daily consumption of bark, twigs, and tree leaves ranged from 1.5 to 2.2 lb (0.7–1.0 kg) for one beaver per day[6] to 4.5 lb (2 kg).[2–4] It should be noted that in addition to bark, beavers eat much herbaceous and aquatic vegetation during the growing season. It is nearly impossible to measure how much of these they take; therefore, most calculations of food intake have been confined to woody plants. Among these, quaking and bigtooth aspen are the most preferred species, even though a great many different trees are utilized, depending on the species composition of the woodlands. Aspen biomass and growth have been well investigated and related to beaver utilization.[5] Therefore, aspen will be used here for caloric estimates.

One kilogram of fresh aspen contains about 2543 kcal (1156 kcal/lb).[7] Even though the beavers eat only the bark, they still ingest much cellulose, as can be seen from the wood fiber–containing feces. However, beavers are able to digest only about 33% of the cellulose taken up.[8] Considering further that overall digestibility decreases as fiber content goes up, it has been estimated that about 50% of a tree's biomass is digestible for a beaver. This includes leaves, twigs, and most of the bark. If we accept this for the calculations, then 1 kg of fresh aspen would provide the beaver with 1272 kcal (580 kcal/lb).

An acre of aspen produces 5840 lb (2651 kg) of food, sufficient energy for 10 beavers for 442 days, assuming they do not grow during that time. If 5840 lb is consumed over 1 year, there would be 2.8 lb/day available for growth for the whole group.

Energy Needs

How many calories does a beaver need per day? Stephenson[9] calculated that a 26-lb (11.8-kg) beaver requires 850 kcal (digestible energy)/day for maintenance, not allowing for growth, and Pearson[10] arrived at a value of 760 kcal. The first figure corresponds to 0.7 kg of fresh aspen. A larger adult, weighing 16–18 kg, would have to consume 0.8–0.9 kg. These estimates agree well with laboratory experiments (0.7 kg/beaver and day[9]) and field observations (0.7–1.0 kg[6]). A beaver needs 1.5 lb (850 kcal) of fresh aspen per day to maintain itself in summer, 3.6 lb (2040 kcal) for maximal growth, and 2.2 lb (3340 kcal) in winter.[6]

Energy Storage

The North American beaver ranges over a vast north-south distance from the Arctic Ocean to Mexico and encounters great contrasts in ecological challenges. For instance, northern beavers in Wood Buffalo National Park (Alberta, Canada) spend about 150 days under the ice each year[11] (Fig. 4.1). By contrast, Ohio beavers are icebound "for no more than a few weeks in January and February."[12]

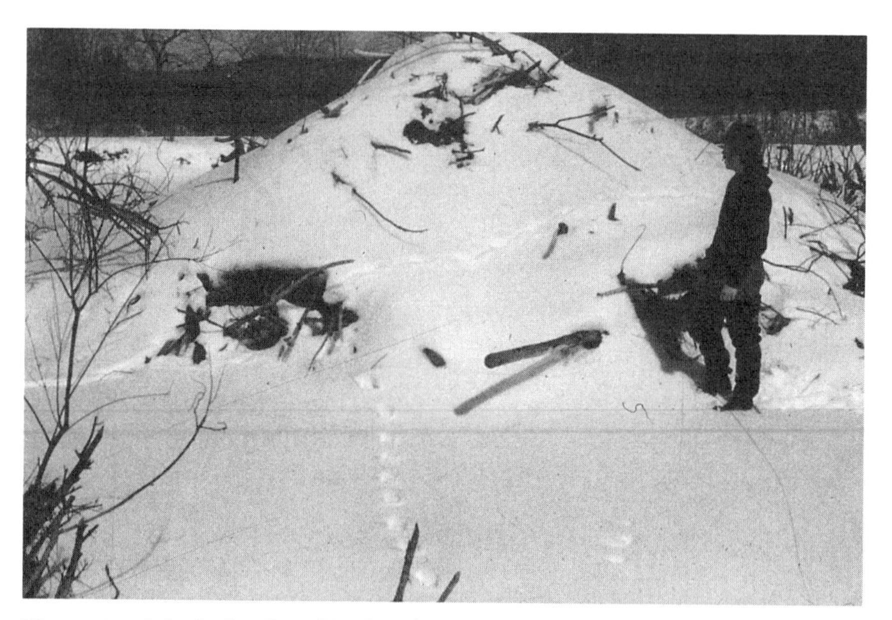

Figure 4.1 | Lodge in winter. Northern beavers are icebound for weeks or months, living in a frozen, snow-covered lodge and reducing their metabolism.

Finally, beavers in the southern United States do not experience snow or ice at all and have access to fresh vegetation year-round.

Beavers store energy for lean times in two forms: body fat and a food cache (Plate 4) near their lodge. Fat accumulates not only in the body parts typical for mammals (intraperitoneal and subcutaneous depots) but also in the tail. From spring to fall, tail volume doubles. As the fat is used up, the dimension of the tail decreases in winter.[13,14] The fat content of the tail grows from about 6.7% in spring and summer to as much as 63.7% in fall and early winter.[14] Loss of up to 34% of body weight can sustain the metabolic rate at 1.9 times the basal metabolism without any ill effect for the animal.

In fall, beavers collect branches and pile them up in the water near their lodge. However, these food caches fail to meet the energy requirements in winter.[11] Hence, adults reduce their body temperature and metabolism and lose weight. Kits, however, gain weight over the winter (see Fig. 10.2).[15] Two- to 4-year-old beavers lose about 5 lb (2.3 kg) over the winter, while 1- to 2-year-olds gain 3–4 lb. The older beavers need to forage or use their stored fat. The fur and the insulation afforded by the lodge help them to conserve energy. The food available for the icebound period of 150 days ranges from 87.6 to 143.1 lb/beaver (average 122.6 lb, or 55.7 kg), or 0.58–0.95 lb/day and beaver. It has to be considered that the stored food is low in protein. Much of the bark contains less than 6% protein. Such food is considered deficient. In Novakowski's samples,[11] protein content averaged 4.39%. Alder and birch provide more protein than aspen and red osier. Beavers also reduce locomotion: on average each animal ventured out from the lodge only once every 2 weeks.

Metabolic Rate and Ambient Temperatures

What environmental conditions force a beaver to burn extra calories? Physical activity and extreme cold increase the metabolism of an animal. Let's look at the effect of temperature, so important for an animal like the beaver that has colonized cold regions in both the Old and the New World. From about 0°–2°C to 28°C, beavers do not change their heat production, which is expressed as watts (W) per kilogram of body weight. To calculate this, one measures the oxygen uptake and carbon dioxide production of an animal. The range from about 0°C to 28°C is called the *thermoneutral zone* of the beaver, meaning an increase or a decrease of ambient temperature has no effect on its metabolic rate. At thermoneutrality, adult and yearling beavers use about 2 W/kg body weight. When fasting, beavers reduce their metabolism (i.e., heat production) to 1.7 W/kg body weight. At temperatures below freezing, metabolic rate increases. At an air temperature of –20°C, adult and yearling beavers metabolize at 3.5 W/kg body weight (instead of only 2.0 W/kg in the thermoneutral zone). Kits have a higher metabolic rate than adults: 5.27 W/kg body weight. By mid-August, the 2-month-old kits have

Table 4.1 | Some Benchmarks Related to Energy Budget of Beavers

Digestive capacity	3458 g wet weight/day
Digestive organ capacity	1484 g wet weight of food
Digestive tract empties	2.33 times/day
Foraging time	269 minutes/day
Energy requirement for adult beaver (15 kg)	1213 kcal/day
Woody biomass harvested per beaver[a]	1400 kg/ha/year
Maximum foraging distance from pond	50–60 m, up to 250 m
Lodge temperature	about 20°C

Source: Reference 18.
[a] From reference 19.

lowered their rate to 3.23 W/kg. This is still 55%–70% higher than in adults. The lower end of the thermoneutral zone (the "lower critical temperature") in a 2- to 3-week-old kit is 24°C, compared with 0°C in adults. By their first autumn, the thermoneutral zone of kits is practically the same as in adults.[16]

Metabolism in Water

A beaver resting in water with a temperature of 19°–31°C has a metabolic rate of 2.4 W/kg body weight. This is 22.4% higher than when resting in air. Beavers in water maximally raise their metabolic rate to 2.2 times that in air.[17]

Table 4.1 lists some benchmarks related to the energy budget of beavers.

REFERENCES

1. Aldous, S. E. 1938. Beaver food utilization studies. Journal of Wildlife Management 2: 215–222.
2. O'Brien, D. F. 1938. A qualitative and quantitative food habit study of beavers in Maine [M.Sc. thesis]. Orono: University of Maine.
3. Warren, E. R. 1927. The beaver: its work and its ways. Baltimore: Williams and Wilkins.
4. Bradt, G. W. 1947. A study of beaver colonies in Michigan. Journal of Mammalogy 19: 139–162.
5. Stegeman, L. C. 1954. The production of aspen and its utilization by beaver on the Huntington Forest. Journal of Wildlife Management 18: 348–358.
6. Brenner, F. J. 1962. Foods consumed by beavers in Crawford County, Pennsylvania. Journal of Wildlife Management 26: 104–107.
7. Cowan, I. McT., W. S. Hoar, and J. Hatter. 1950. The effect of forest succession upon the

quantity and upon the nutritive values of woody plants used as food by moose. Canadian Journal of Research 28: 249–271.

8. Currier, A., W. D. Kitts, and I. McT. Cowan. 1960. Cellulose digestion in the beaver (*Castor canadensis*). Canadian Journal of Zoology 38: 1109–1116.
9. Stephenson, A. B. 1956. Preliminary studies on growth, nutrition and blood chemistry of beavers [M.S. thesis]. Vancouver: University of British Columbia.
10. Pearson, A. M. 1960. A study of the growth and reproduction of the beaver (*Castor canadensis* Kuhl) correlated with the quality and quantity of some habitat factors [M.S. thesis]. Vancouver: University British Columbia.
11. Novakowski, N. S. 1967. The winter bioenergetics of a beaver population in northern latitudes. Canadian Journal of Zoology 45: 1107–1118.
12. Svendsen, G. E. 1980. Seasonal change in feeding patterns of beaver in southeastern Ohio. Journal of Wildlife Management 44: 285–290.
13. Patric, E. F., and W. L. Webb. 1960. An evaluation of three age determination criteria in live beavers. Journal of Wildlife Management 24: 37–44.
14. Aleksiuk, M. 1970. The function of the tail as a fat storage depot in the beaver (*Castor canadensis*). Journal of Mammalogy 51: 145–148.
15. Smith, D. W., R. O. Peterson, T. T. Drummer, and D. S. Sheputis. 1991. Over-winter activity and body temperature patterns in northern beavers. Canadian Journal of Zoology 69: 2178–2182.
16. MacArthur, R. A. 1989. Energy metabolism and thermoregulation of beaver (*Castor canadensis*). Canadian Journal of Zoology 67: 651–657.
17. MacArthur, R. A., and A. P. Dyck. 1990. Aquatic thermoregulation of captive and free-ranging beavers (*Castor canadensis*). Canadian Journal of Zoology 68: 2409–2416.
18. Belovski, G. E. 1984. Summer diet optimization by beaver. The American Midland Naturalist 111: 209–222.
19. Johnston, C. A., and R. J. Naiman. 1990. Browse selection by beaver: effects on riparian forest composition. Canadian Journal of Forest Research 20: 1036–1043.

Part II

Behavior

chapter 5

Families as Social Units

It is a peculiarity of the languages of our Indian nations that . . . they are opulent in terms for the designation of natural objects, and for expressing relative differences in the same object. In the Ojibwa, for example, there are different names for the beaver according to his age, and compound terms to indicate sex, as follows: Specific name, Ah-mik'. Year old and under, Ah-wa-ne-sha'. Two years old, O-bo-ye-wa'. Full grown, or old, Gî-chî-ah'-mik. Male beaver, Ah-yä-ba-mik'. Female beaver, No-zha-mik'.

Lewis H. Morgan, 1868

Family Size and Composition

The basic social unit of beaver society is the family. It consists of the parents, the young of the year, and yearlings. Two-year-olds may or may not be present. They usually leave or are expelled when or before a new litter is born. On average, there are 2 kits of the year and 2 yearlings, so that a typical family on a reasonably good site numbers about 6 members, provided there is no trapping or other disturbance to the colony. Family size typically ranges from a kitless pair (or even a temporarily single adult) to about 10 members. Of 46 colonies in Newfoundland, 7 (15%) consisted of single individuals, 11 (24%) were pairs, and 28 (61%) entire families.[1] Families rarely grow to more than 15 members. In our study area, Allegany State Park, New York, families of 9–10 beavers are common. In different years, the average family size varied from 4.1 to 6.12 individuals.[2] In Ohio, the average family size was 5.85 before the older young dispersed, and 6.04 after new kits were born.[3]

Two-year-olds may stay on for another year. This happens more often when the population density is high. The average percentage of colonies with such an "additional adult" was 38% in five studies. It ranged from 22% to 87.5%. On average, only 5.5% of families in populations at lower densities (in five studies), mostly due to trapping, had an "additional adult" (i.e., 2- to 3-year-old), with a range of 0%–12.5%.[2]

Larger families build additional lodges and expand their feeding range. Families grow large where beavers are provisioned and otherwise protected, as for instance on the Beaversprite Sanctuary in Upstate New York. The size of the food cache does not reflect colony size.[4]

Beavers are monogamous. Recent genetic studies on a variety of mammals and notably birds[5] showed that we must distinguish between *social* and *genetic* monogamy. While a particular pair may live together year after year, so-called extra-pair copulations are possible. Only genetic analysis will reveal the parentage of the offspring. In the case of the beaver, future studies will show whether its monogamy is social as well as genetic.

Lodge Use and Site Defense

All family members might inhabit one lodge. This is especially true during the winter months when only one lodge is provisioned with a food cache. During the summer months, several lodges may be used (Plate 5). Often the kits are moved to an auxiliary lodge, or the adult male occupies a separate lodge for several weeks while the young are still small and being nursed. When the mother tries to wean the young, she may separate herself from them by living in another lodge within the colony.

Home Range

The family lives on a particular site year-round and defends it against conspecifics at all times. Such site tenacity permits all members of the colony to become thoroughly familiar with all important parts of their home range. Traditionally, a defended area is termed *territory*, while the entire area that an animal or group uses is the *home range*. For beavers, home range and territory may not be too different, as they defend the ponds they have created by habitat modification. Nevertheless we can consider the feeding areas more distant from the water as being home range, while the territory encompasses the impoundments with lodges, dams, canals, trails (Plate 6), and food caches. This center of the home range is also known as the *core area*.

Home ranges have been measured as a linear distance along streams, or as an area. Some examples are listed in Table 5.1. Once settled down, beavers rarely move from their pond, especially when they are caring for helpless newborn kits inside the lodge. They often stay in the same location for several years. However, beavers sometimes are forced to relocate, as when an unexpected flood washes out their dam. Each adult then moves a kit to a safe place by carrying it between his or her own chin and the two forepaws. The parents transport all kits, one by one. Renee, one of our dedicated field assistants, was lucky to be the only one to witness this spectacle of a beaver relocating its kit.

Table 5.1 | Sizes of Home Ranges of Beaver Families in Several North American Areas

Geographical Area	Size in Length or Area
Manitoba (tundra)[6]	2.3–42.8 ha
Ontario[7]	1.8 ha in ponds–15 ha in lakes
California[8]	200–800 m along stream
South Carolina[9]	Extreme points within 1 colony in 1 year 84–1863 m apart
South Carolina[10]	6.7 ha (pond)–16.13 ha (creek)

Behavior inside the Lodge

For obvious reasons, behavior inside the beaver lodge has remained the last bastion resisting observation and analysis. Four modern techniques have provided us with more than glimpses into this most elusive part of the beaver's life. These methods include continuous recording with microphones,[11,12] direct observation and filming from a blind attached to a lodge,[13] radiotelemetry,[14,15] and videorecording via a spy camera inserted into the lodge, pioneered by Dr. Donald Griffin.[16]

These studies shed light on events that occur around the time of birth. When the female gives birth, the male often moves to another lodge on the same site. The eyes of the newborn open within minutes[17]; this had been a matter of dispute. Inside the lodge, the kits play with each other infrequently, but rarely with the yearlings or parents.[18] Griffin recorded "housecleaning" behavior in great detail: the beavers push out old, decaying woody material and dump it into the plungehole of the lodge. Beavers even seem to "wash their bedding": When the kits still spend all their time in the lodge, mostly the yearlings were observed to clean the floor of the chamber every 2–3 days. They pushed the bedding into the water and back up into the chamber several hours later. The male and female collected fresh plant material and added it to the litter.[18]

Surprisingly, many other animals inhabit beaver lodges. Patenaude[17,18] documented muskrats scavenging among the beavers in the lodge, whereas Griffin[16] filmed voles, mice, and moths sharing the living chamber.

The 2-year-olds proved especially fascinating in the study by Patenaude[17] and her coworkers in Quebec. These researchers opened up a beaver lodge and recorded the behavior of the inhabitants with a video camera. They observed that these older young assist their parents in raising younger siblings. They feed, groom, and guard the newborn kits. Although these "helpers" thus contribute to the parents' success in raising another litter, they postpone their own breeding attempts for at least 1 year. So instead of having offspring by themselves, they indirectly increase their own reproductive success by boosting the survival chance of their siblings. This choice is particularly adaptive under conditions of food shortage, of high population density with a shortage of potential sites for new breeding pairs, or during catastrophes such as a drought that dries out many beaver ponds. Under these adverse conditions, dispersal would be less likely to succeed. The dispersers would die instead. Therefore, these 2-year-olds are better off by staying on and helping their parents with housekeeping and baby-sitting their younger siblings.

The beaver is not alone in practicing such "cooperative breeding." Similar circumstances such as a saturated habitat lead to cooperative breeding in other vertebrates, notably birds such as crows, jays, and woodpeckers.

Parenting

The mother gives birth to one to several kits every year in May or June. In a lodge in Gatineau Park, Quebec, monitored through a window of an attached observation hut, the female, male, and one yearling "formed a triangle around the newborn."[18] The female licks the kits thoroughly after birth and occasionally for 2 or 3 more days. Only the mother ingests the excrement of the young during the first 2–3 days, until the kits defecate and urinate in the water inside the lodge, enjoying a veritable "indoor water toilet." After nursing the kits for about 2 weeks, the mother hauls tender tree branches into the lodge for the kits to try (Plate 7). The kits stay in the lodge until they are about 1 month old. Their first excursions outside the lodge are not necessarily voluntary. In Griffin's video,[16] parent beavers push kits down the plungehole despite their vocal protest. Clinging to the parent's fur during an excursion outside the lodge is another way of resolving the conflicting efforts by the two generations. During the first days out in the water, the kits stay near their parents and the lodge and seek instant refuge when danger threatens. During these explorations kits still do not forage for themselves; parents and older siblings deliver leafy twigs to them until they are 2–3 months old. Like the young of other mammals, beaver kits are alarmed easily. They often slap their tails on the water, normally an alarm signal. But adults do not respond to these "false alarms." In contrast, the parents' tail slaps invariably trigger escape by diving. The kits whine when they compete for food the parents bring into the lodge. The same sound is heard when kits and other young beavers compete for choice food outside the lodge, either among themselves or with the parents. Weaning occurs at 10 weeks.[17] After that time, the young can be seen feeding side by side on branches brought into the pond by parents and older siblings (Plate 8).

In fall, adults and yearlings build the food cache. During winter, mostly the adults fetch branches from the food pile and rhizomes from the lake or pond bottom for the entire family. The yearlings participate less frequently in this work.[18]

Beaver parents are constantly working on one project or another, literally being "busy as a beaver." They repair dams, fortify their lodges, clean out dens, dig canals, cut down trees, scent-mark, patrol their pond, feed and preen themselves, and bring leafy twigs for the kits into the lodge. Older offspring often help their parents. However, beavers less than 1 year old play and explore much, although they also copy their parents' behavior, albeit in an incomplete form. For instance, we observed yearlings imitating their dam-building parents: While the parents dredged up mud from the pond bottom near the dam and added it to the dam, a youngster also dived and picked up mud, but much farther from the dam. Mud in paws, the young beaver swam to the dam, leaving a wake of muddy water. He arrived at the dam empty-handed.

Recognition of Family Members

Coordination of social behavior, separation of roles, and defense against invading beavers require that beavers recognize family members. Fortunately, beavers possess an extremely acute recognition system based on chemical secretions. The anal gland secretion can be used to match the body odors of other individuals to distinguish family members from nonmembers ("phenotype matching"). Beavers also memorize the peculiar odor of castoreum of family members so that all members of a colony can be identified.

In fact, beavers can even recognize former family members or other close relatives after a long absence, and even if they have never met before, they can recognize a sibling born after they have departed from the parental lodge. This is probably accomplished through phenotype matching using the anal gland secretion. We recorded in great detail the identities and movements of all beavers in two neighboring families, about 2 km apart. The adult male of the upper-stream colony had dispersed from the downstream colony several years ago. In the summer of 1994 we were surprised to find that one of the sons of the upper-stream colony stayed in the downstream colony for about 2 months before he continued his journey of dispersal. If this beaver had not been the grandchild of that downstream pair, they would not have admitted him but would have attacked him instead.

Division of Labor: Sex Differences

Do the sexes play different roles in the social life of the colony? To answer this question, we first have to be able to distinguish male and female beavers. This is not easy, as the sexes are alike in size and physical appearance, and hence indistinguishable to even the most experienced field observer. Lack of external sex differences is typical for monogamous species of mammals such as some primates, canids and rodents, and birds such as penguins and terns. Reduced sexual dimorphism, monogamy, and the male's participation in rearing the young go hand in hand.

Sex Determination

From a distance, the most obvious sex characteristic in beavers is the teats of the lactating female. This, however, is limited to a brief period in summer when the young suckle. Even then, some females do not breed and will not develop conspicuous teats. Therefore, this method can't be used to reliably separate male and female beavers. The other methods require handling an animal. First, a male can be distinguished by palpation of the abdominal region. Here the cartilaginous baculum ("penis bone") can be found. It feels like a bean. Second, the color and viscosity of the anal gland secretion differs between the sexes. In the North American beaver, males have darker, brownish, and more viscous secretion, while the female secretion is whitish, cream-colored, and more liquid.[19] In the Eurasian beaver, the anal gland secretion in the male is a yellowish fluid, while in the female it is a grayish paste.[20] Third, the sexes differ in number and arrangement of the openings in the cloaca. The fourth, very reliable method of sexing beavers is to determine chromosomal sex differences. For beavers, blood smears are taken from a hind toe and are stained, and the polymorphonuclear leukocytes examined under the microscope.[21]

Role Differences of Sexes in Family Life

The adult female plays a pivotal role in the social life of a beaver colony. In a study in the northeastern United States, she emerged first from the lodge in the evening, warned colony members more often by tailslapping, was most dominant in encounters, and led the family in maintaining the dam and lodge and building food caches. The male, on the other hand, was more active in inspecting the dams closely.[22]

The results of several studies varied with regard to the respective roles of the sexes. In northern Minnesota the female provided more energy by nursing and feeding the young, while the male primarily built and maintained structures and showed more alarm behavior. The female scent-marked only from May through July, while the male did so all summer long but with decreasing intensity. In several studies the male scent-marked more than the female.[23–25] The male spent more time in and near the lodge when kits of the year were present. He probably provided parental care. After several weeks of nursing, the female avoided her kits by moving to an accessory lodge, thus probably weaning the young by force.[25]

Finally, no modern researcher has found a beaver family even nearly the size depicted in the delightful etching, filled with artistic license, of a colony at Niagara Falls, dating from the early 18th century (Fig. 5.1).

Figure 5.1 | Artistic license: over 50 beavers in one colony near Niagara Falls, carrying building materials in preposterous ways. Early 18th century. (From: G. Pilleri, editor. 1984. Investigations on beavers. Volume 3. Berne: Brain Anatomy Institute. Reproduced with permission from the British Library.)

REFERENCES

1. Payne, N. F. 1982. Colony size, age, and sex structure of Newfoundland beaver. Journal of Wildlife Management 46: 655–661.
2. Müller-Schwarze, D., and B. A. Schulte. 1999. Behavioral and ecological characteristics of a "climax" population of beaver (*Castor canadensis*). In: P. E. Busher and R. M. Dzięciołowski, editors. Beaver protection, management, and utilization in Europe and North America. New York: Kluwer Academic/Plenum. p 161–177.
3. Svendsen, G. E. 1980. Population parameters and colony composition of beaver (*Castor canadensis*) in southeast Ohio. American Midland Naturalist 104: 47–56.

4. Osmundson, C. L., and S. W. Buskirk. 1993. Size of food caches as a predictor of beaver colony size. Wildlife Society Bulletin 21: 64–68.
5. Westneat, D. F. and P. W. Sherman. 1997. Density and extra-pair fertilizations in birds: a comparative study. Behavioral Ecology and Sociobiology 41: 205–215.
6. Wheatley, M. 1997. Beaver, *Castor canadensis*, home range size and pattern of use in the taiga of southeastern Manitoba: I. Seasonal variation. Canadian Field Naturalist 111: 204–210.
7. Gillespie, P. S. 1976. Summer activities, home range and habitat use of beavers [M.S. thesis]. Toronto: University of Toronto.
8. Busher, P. E. 1975. Movements and activities of beavers, *Castor canadensis*, on Sagehen Creek, California [M.A. thesis]. San Francisco: San Francisco State University.
9. Davis, J. R., A. F. Von Recum, D. D. Smith, and D. C. Guynn Jr. 1984. Implantable telemetry in beaver. Wildlife Society Bulletin 12: 322–324.
10. Boller, L. J., and T. T. Fendley. 1989. Beaver home ranges in the South Carolina Piedmont [abstract]. 5th International Theriological Conference; 22–29 August 1989; Rome, 3rd International Beaver Symposium. p 296–297.
11. Bovet, J. and E. F. Oertli. 1974. Free-running circadian activity rhythms in free-living beaver (*Castor canadensis*). Journal of Comparative Physiology 92: 1–10.
12. Potvin, C. L., and J. Bovet. 1975. Annual cycle of patterns of activity rhythms in beaver colonies (*Castor canadensis*). Journal of Comparative Physiology A Sensory Neural and Behavioral Physiology 98: 243–256.
13. Patenaude-Pilote, F., E. O. Oertli, and J. Bovet. 1980. A device for observing wild beavers in their lodge. Canadian Journal of Zoology 58: 1210–1212.
14. Lancia, R. A., and W. E. Dodge. 1977. A telemetry system for continuously recording lodge use, nocturnal and subnivean activity of beaver (*Castor canadensis*). In: F. M. Long, editor. Proceedings First International Conference on Wildlife Biotelemetry, University of Wyoming, Laramie. Laramie: First International Conference of Wildlife Biotelemetry.
15. Sokolov, V. E., V. A. Rodionov, V. P. Sukhov, V. S. Kudryashov, and M. S. Kuznetzov. 1977. A radiotelemetrical study of diurnal activity in *Castor fiber*. Zoologícheskii Zhurnal, 56: 1372–1380 [in Russian with English summary].
16. Griffin, D. 1999. Unpublished film.
17. Patenaude, F. 1982. Une année dans la vie du castor. Les Carnets de Zoologie 42: 5–12.
18. Patenaude, F. 1983. Care of young in a family of wild beavers, *Castor canadensis*. Acta Zoologica. Fennica 174: 121–122.
19. Schulte, B. A., D. Müller-Schwarze, and L. Sun. 1995. Using anal gland secretion to determine sex in beaver. Journal of Wildlife Management 59: 614–618.
20. Rosell, F., and L. Sun. 1999. Use of anal gland secretion to distinguish the two beaver species *Castor canadensis* and *C. fiber*. Wildlife Biology 5: 119–123.
21. Larson, J. S., and S. J. Knapp. 1971. Sexual dimorphism in beaver neutrophils. Journal of Mammalogy 52: 212–215.
22. Hodgdon, H., and J. S. Larson. 1973. Some sexual differences in behaviour within a colony of marked beavers (*Castor canadensis*). Animal Behaviour 21: 147–152.
23. Hodgdon, H. E. 1978. Social dynamics and behavior within an unexploited beaver

(*Castor canadensis*) population [Ph.D. dissertation]. Amherst: University of Massachusetts.

24. Svendsen, G. E. 1980. Patterns of scent-mounding in a population of beaver (*Castor canadensis*). Journal of Chemical Ecology 6: 133–149.
25. Buech, R. R. 1995. Sex differences in behavior of beavers living in near-boreal lake habitat. Canadian Journal of Zoology 73: 2122–2143.

Communication by Scent and Sound

6 chapter

When two Beaver lodges are in the vicinity of each other, the animals proceed from one of them at night to a certain spot, deposit their castoreum, and then return to their lodge. The Beavers in the other lodge, scenting this, repair to the same spot, cover it over with earth, and then make a similar deposit on the top. This operation is repeated by each party alternately, until quite a mound is raised, sometimes to the height of four or five feet.

J. J. Audubon and J. Bachman, 1854

And if a Woman with childe go over a *beaver,* she will suffer abortment; and *Hippocrates* affirmeth, that a perfume made with *Castoreum,* Asses dung, and Swines grease, openeth a closed womb.

Edward Topsell, 1658

Water covered the Earth in the beginning. Beaver, otter and muskrat lived in this water. They dived and brought up mud. The Great Spirit Manitou created the dry land from this mud.

Creation myth of the Amikonas ("People of the Beaver"), a Native American tribe near Lake Huron

Scent Mounds and Scent-Marking Behavior

To this day, beavers keep dredging up mud from the bottoms of their ponds, and not only for their dams and lodges. Like most mammals, beavers frequently communicate with one another by chemical signals. The most conspicuous sign of scent marking is the scent mound (Plate 9). From pond sediment, the beaver builds a mud pile on which it places its territorial marks. This is unique among mammals. How does the beaver construct a scent mound? First the beaver dives to the bottom of the pond, dredges up a batch of mud with its hands, and holds it against its chest. Carrying this load, the beaver emerges and waddles on its hind feet. After a few steps, it deposits the mud on land, close to the bank. The beaver then arches its back, straddles the mud pile, and applies castoreum or anal gland secretions, or both, through the cloacal opening. From close by, one can hear a noise as the beaver expels the castoreum. A mud pile can consist of just one "load" or may measure up to 50 cm in diameter and 30 cm in height. Such a "giant" scent mound is shown in Plate 9. Sometimes beavers repeatedly mark the same spot and add more and more mud in the process. The largest one we have ever measured was about 2.5 feet (80 cm) in diameter and 20 inches (50 cm) high, found at Cranberry Lake in New York's Adirondack Mountains. Beavers also scent-mark without using mud. Often they twist a bunch of grass and then squirt it with secretions.

Why do beavers build a mound first before they spray it with castoreum? There are at least three reasons. The first is to elevate the point of odor release. Chemical ecologists say the *active space* is increased this way. The active space is the volume of air downstream from the odor source where the concentration of the odor is strong enough to be noticed by other beavers. Second, a moist substrate like mud helps to intensify the odor. Many of us have experienced intensified odors, such as car exhaust fumes, on rainy days, especially when the rain starts. Likewise, a wet dog has a more intense odor, and dogs themselves track better on humid days. On a wet surface, water and odor molecules compete for surface sites. The more water vapor there is, the more odor molecules are released from the surface of the scent mound and reach the mucus on the olfactory epithelium in the nose. Furthermore, when dry solid material dissolves in water, it can volatilize better. Finally, the mound protects this now raised odor beacon from flooding when the water level at the beaver pond fluctuates.

Beavers build most of their scent mounds in spring, especially in April.[1,2] It is the time when dispersers are abundant and try to seize suitable but undefended sites. Marking abates in summer and fall when invasion pressure has declined. However, both Eurasian beavers in Norway and North American beavers in New York sometimes scent-mark on snow. Older beavers scent-mark more frequently than the younger members of their families. The adult male in a family contributes most to building and marking scent mounds.[3]

To maximize the effect of their signals, beavers deposit their scent mounds at some strategic locations, often near the paths most likely used by invading beavers.

Functions of Scent Marks

The function of the castoreum marks on the scent mounds has been investigated in several field studies in the eastern United States. In 1968 Aleksiuk[4] postulated that the scent mounds served in warning off transient beavers from areas that are occupied. This is the reason for building scent mounds in spring when many 2-year-olds begin to disperse and many beaver families suffer a high rate of

Figure 6.1 | An experimental scent mound (upper right) destroyed by pawing and marking. At left, an undisturbed experimental scent mound.

invasion. To confirm this territorial function, we tricked unsuspecting beavers by placing beaver scent mounds, and also castoreum from beavers of other families, in a territory to observe the response of resident beavers.[17] They typically responded to an alien scent mound by sniffing it first (Plate 10), then pawing it with their front feet, and finally straddling and marking it (Plate 11). Often they destroyed mounds that didn't smell like their family members (Fig. 6.1). Occasionally, they built one or several new mounds near our experimental mound, presumably to "out-stink" the alien scent. They even picked up the experimental mud and incorporated it into their own mound. When we applied castoreum samples on corks, the resident beavers often chewed the cork into pieces.

Scent Glands and Their Secretions

Castoreum from Castor Sacs

Beavers possess a pair of unique pouchlike structures called *castor sacs* (Plate 12). They contain the yellowish fluid that is the *castoreum*. The two castor sacs are located between the kidneys and urine bladder, opening into the urethra. They are large, weighing about 60 g, which represents about 0.3% of the body weight of an average adult beaver of 19 kg. They are slightly (9 g) smaller in females than in males,[5] although Svendsen[6] found no sex differences. The castor sacs contain connective tissue but no secretory cells.[6]

Chemical Composition The chemical composition of castoreum was investigated early because of the intense interest of perfumers. There are many compounds, belonging to many different classes, and about 45 of them have been identified.[7] A particularly interesting compound is castoramine.[7,8] Related compounds are found in pond lilies (genus *Nuphar*). Beavers feed on these lilies, and the metabolic pathways of these "Nuphar alkaloids" are of considerable interest at the present time. The other compounds belong to a great variety of classes such as phenolics, terpenes, alcohols, aldehydes, ketones, esters, and carboxylic acids.

Many of the castoreum compounds can also be found in the beavers' food plants (Table 6.1). Therefore, castoreum seems to be derived from the beaver's diet. How beavers actually accumulate and concentrate castoreum compounds in the castor sacs is still a mystery.

A close look at the castoreum constituents reveals that quite a few are used by plants to defend against herbivores such as insects and mammals, which can damage plants considerably. These antifeedants are called *secondary compounds* because they are not part of the main metabolism of carbohydrates, fats, and proteins; are not essential for growth and nutrition; and are not common to many organisms. Secondary compounds are specific to certain plant species and defend against microbes and larger herbivores. Such secondary compounds have many different effects on animals, ranging from interrupting digestion to inhibiting reproduction to being outright toxic and deadly. While natural selection has forced many plants to resist herbivores by strong chemical defenses, herbivores in turn have developed many ingenious ways to deal with or circumvent the effects of these plant secondary metabolites. Plants and herbivores are locked in an arms race: measure prompts countermeasure. As obligatory, albeit generalist herbi-

Table 6.1 | Castoreum Compounds That Are Also Found in Beaver Food Plants

Compound Found in Castoreum	Beaver Food with Same Compound
Benzoic acid	Black cherry, Scots pine
Benzyl alcohol	Aspens, poplar trees
Borneol	Bark beetle–infested pines
Catechol	Common cottonwood
o-Cresol	Northern white cedar
4-Ethyl phenol	Norway spruce, poplar trees
4-(4'hydroxyphenyl)-2-butanone	Birches
Hydroquinone	Red pine
Phenol	Scots pine
Salicylaldehyde	Willows, poplar trees

Source: Modified from reference 9.

vores, beavers refined one of these methods: they sequester the potentially adverse plant compounds and recycle them as signals for territorial advertisement.

Beavers are not the only ones who use castoreum. For millennia humans have appreciated the medical and perfume effects of castoreum. The Romans burned it in lamps and had women inhale the fumes because they believed this would induce abortions. Medieval beekeepers treated honeybees with castoreum to increase honey production and to keep predatory bees away. According to Linnaeus, Lapplanders mixed castoreum with snuff.[10] Until the 1700s castoreum was prescribed for many maladies, from headaches to fever and hysteria. Homeopathic handbooks in Europe still list castoreum, but it is rarely used nowadays. Castoreum contains salicylic acid and salicylaldehyde. Salicylic acid is the active ingredient in aspirin. As the name says, it originates in willow, scientifically the genus *Salix*. Native Americans used willow bark against headaches. Did they watch beavers chewing large amounts of willow bark and try it themselves? Truly a fountain of health, salicylic acid serves as an antiseptic, disinfectant, analgesic, and antipyretic (fever-fighting), and against rheumatism, acne, and warts.

We have collected castoreum, in its fresh state better called *castor fluid*, for behavioral experiments and for chemical analysis. "Milking" live, immobilized beavers yields copious amounts (Plate 13).

Behavioral Effects of Constituents By presenting castoreum components one by one to free-living beavers and observing their responses, we sifted out 15 compounds that potentially can elicit resident beavers' territorial behavior. These odors attracted beavers, who often destroyed the marks and "overmarked" them. The different compounds carry the same message of territorial claim by outsiders, which cannot be tolerated by a territory holder. They serve the same territorial function and are thus redundant, because one compound should be enough to deliver the message. Why do so many compounds convey the same information? It seems one way to ensure that invading beavers get the message without error. In addition, castoreum appears to vary with changing diets, and some compounds may be absent in castoreum at certain times. These redundant odor components are not unlike the several synonyms for one meaning we use in our verbal language to avoid misunderstandings.

Other Functions Several other functions have been suggested for castoreum, although the territorial one may be primary. Butler and Butler[11] proposed that beaver scent mounds provide information about individuals and physiological status within a family. Svendsen[2] agreed with the territorial function but further proposed that scent mounds enhance the confidence of resident beavers and lower that of intruders at the same time. Schulte[12] found that beavers can use castoreum to tell family members from nonmembers and distinguish neighbors from a complete stranger (Fig. 6.2). While these results and hypotheses need fur-

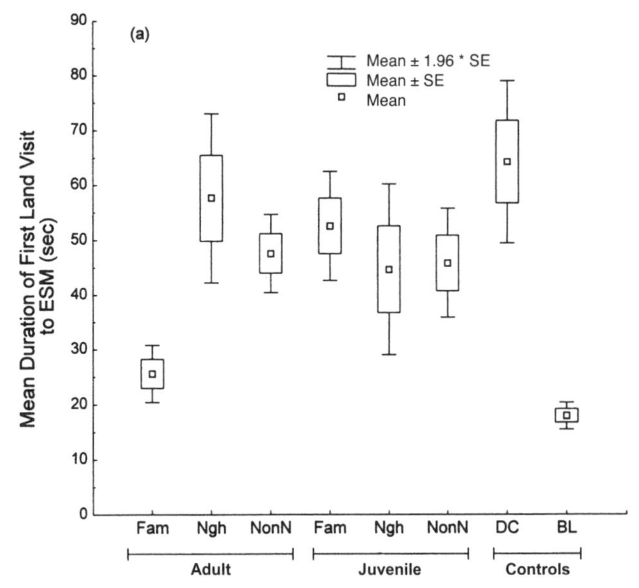

Figure 6.2 | Responses of family members, neighbors, and strangers to castor fluid from adult and juvenile males. In both age classes of odor donors, castor fluid from family members (Fam), neighbors (Ngh), and distant nonneighbors (NonN) were tested. The mean duration of the first land visit reflects the intensity of the responding animal's behavior at the experimental scent mound (ESM). Note that responses to secretion from adult male from the same family is as low as that to blanks (BL). At the other extreme, commercial dried castoreum (DC) triggers strong responses. (Reproduced with permission from B. A. Schulte. 1998. Scent marking and responses to male castor fluid by beavers. Journal of Mammalogy 79: 197, Fig 1a.)

ther confirmation, there is no doubt that castoreum signals more than just territorial occupancy. Decoding the chemical information in castoreum remains a challenge for the future.

Anal Gland Secretion

Species and Sex Differences Beavers have small eyes and ears but an extremely sensitive nose. In encounters with other beavers, their powerful nose helps recognize their sex and discriminate family members and relatives. How can they do it?

A pair of anal glands (Plate 12) discharge their secretion into the "cloaca," the common chamber where also digestive, urinary, and genital tracts terminate.*

*To be anatomically correct: Mammals have lost the typical vertebrate cloaca. The beaver has secondarily acquired a common chamber for the various excretions and secretions. Hence, "cloaca" is put in quotes here.

These organs consist of sebaceous glands and produce an oily secretion. Therefore, trappers speak of "oil sacs." The secretion's color and viscosity can be used to tell the sex of a beaver. In the North American beaver, secretions from males are sticky and brownish, while those from females are whitish/yellowish and watery.[13] To humans, female secretions have a stronger, unpleasant smell, resembling rancid fat. In the Eurasian beaver, secretion from males is yellowish and watery, like that from female North American beavers. By contrast, female secretions are gray and extremely sticky. Both smell rancid, with female secretion having a hint of gasoline odor.[14]

Chemical Composition The anal gland secretion contains tens or possibly over a 100 different chemical compounds in one sample.[15] So far 43 compounds have been found in males only and 67, in females only. Thirty-three compounds occur in both sexes but in an amount characteristic for each sex. In the Eurasian beaver, Grønneberg and Lee[16] identified as main components wax esters in the male but fatty acids in the female. The chemical compounds in the anal gland secretion of the North American beaver have yet to be identified.

Kin Recognition and Other Functions Beavers rub the oily material from the anal gland into their fur to waterproof it, much in the manner birds use their "preen gland." This secretion serves other functions, too. Døving (personal communication, 1987), and more recently, Rosell and Bergan[17] in Norway have observed and videotaped beavers protruding their anal glands and rubbing them on scent mounds. This produces a scent mark of great significance to the beaver.

Beavers' keen olfactory sense is astounding. In a field experiment, we simultaneously presented to beavers anal gland secretions from two different individuals, one their unfamiliar sibling, the other a complete stranger from a distant site (Fig. 6.3). Beavers consistently responded more strongly to the stranger's secretion than to their sibling's. By "more strongly," we mean they were more likely to destroy (by pawing) and overmark the experimental scent-mark (Fig. 6.4). We concluded from this that beavers can recognize their siblings by anal gland secretion only. In our particular study, the test beavers were almost certainly not familiar with either of the secretion donors. Thus, without ever having encountered their siblings before, they can recognize their signs. An example may explain the rationale. Say you have a sibling you have never met. One day, you meet in the street a person who looks very much like yourself. You suspect this person is your sib. You base your judgment of the relationship between the two of you on visual cues signaling similarity. Assessing similarity in this way, resulting in recognition, is called *phenotype matching*. Instead of visual cues, beavers use the chemical composition of anal gland secretion for phenotype matching to determine relationships between individuals. This composition of the secretion resembles a fingerprint: it does not change with changes in diet, location, season, or age.

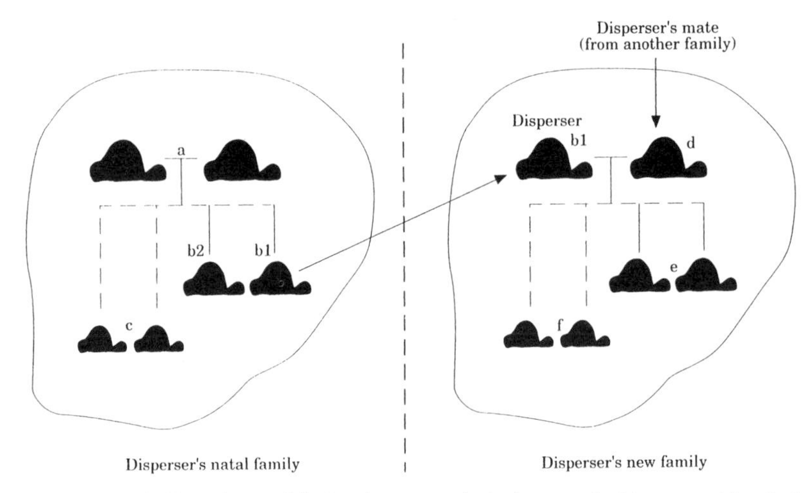

Figure 6.3 | Experimental design (two-way choice) to test for kin recognition in free-living beavers. Scent from members of the colony of origin (on left) is presented to dispersed beavers (on right). The odor donors (c) were born after the disperser had left his or her natal colony. (From L. Sun and D. Müller-Schwarze. 1997. Sibling recognition in the beaver: a field test for phenotype-matching. Animal Behavior 54: 492–502. Reproduced with permission from Academic Press.)

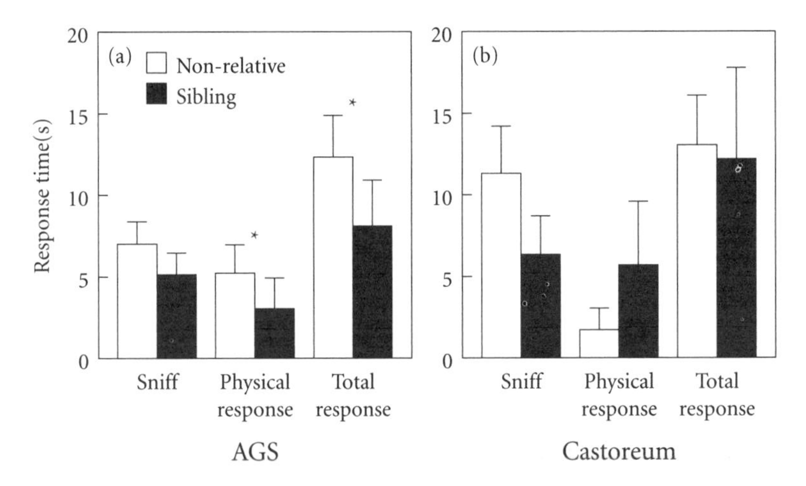

Figure 6.4 | Results of kin recognition experiment. Responses by dispersed beavers to anal gland secretion (AGS) are represented on the left (a) and those to castoreum, on the right (b). "Physical response" refers to pawing and overmarking. Note that dispersed beavers responded more strongly to AGS from nonrelatives than from siblings (marked by asterisk). Castoreum had no such effect. (From L. Sun and D. Müller-Schwarze. 1997. Sibling recognition in the beaver: a field test for phenotype-matching. Animal Behavior 54: 492–502. Reproduced with permission from Academic Press.)

Therefore, to other beavers each individual emanates a unique smell, its olfactory identity card. Moreover, the chemical mix in the anal gland secretions is more similar in related beavers than in nonrelated individuals. In short, beavers use their extremely sensitive nose to tell relatives from nonrelatives by comparing a smell with their own.[18]

Anal gland secretion serves other levels of recognition. It can be used in recognizing sex and family membership.[19] We have just begun to decipher the silent and rich chemical language of beavers.

Population Density and Scent Marking

The number of beaver colonies in a stream system and the number of scent mounds at any one beaver site are correlated: the higher the population density, the more scent marks. This is true for the North American beaver[20] (Fig. 6.5) and the Eurasian species.[21]

By placing experimental scent mounds into suitable but unoccupied sites, it is possible to manipulate colonization by beavers looking for a home. In spring, we treated 25 vacant beaver sites in Fulton County (southern Adirondacks), each with 6 scent mounds containing castoreum and anal gland secretion, and another 25 sites with unscented, control mud piles. By the end of the summer, we recorded whether the site became occupied or not. This was repeated a second year,

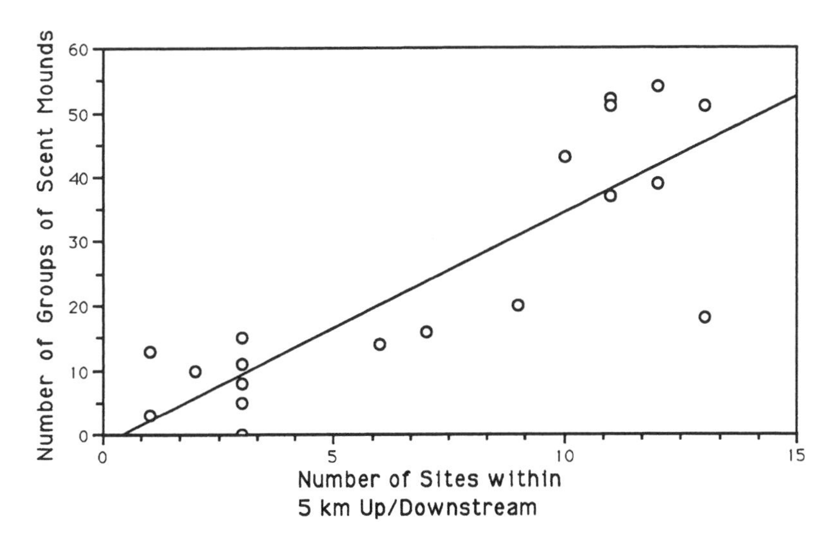

Figure 6.5 | Correlation of population density (number of beaver colonies within 5 km upstream and downstream of a given beaver colony) and number of scent mounds at a given beaver colony. (Reproduced from P. W. Houlihan. 1989. Scent mounding by beaver (*Castor canadensis*): functional and semiochemical aspects [M.S. thesis]. Syracuse: State University of New York College of Environmental Science and Forestry.)

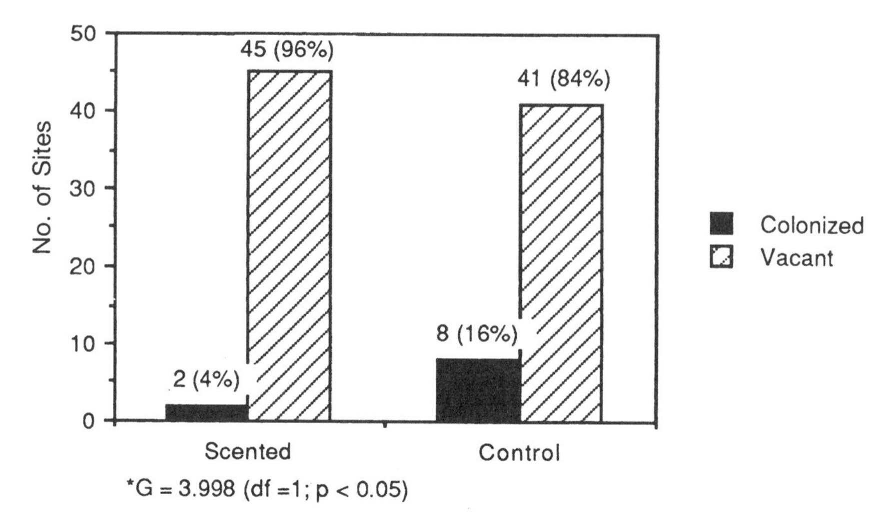

Figure 6.6 | Numbers of vacant sites colonized after experimental scenting (Fulton County Experiment). (From D. Müller-Schwarze. 1990. Leading them by their noses: animal and plant odours for managing vertebrates. Chemical Signals in Vertebrates 5: 591, Fig. 1. Reproduced with permission from Kluwer Academic/Plenum.)

with experimental and control sites swapped. At the end of the experiment, 4 (16%) of the control sites had been colonized, versus only 1 (4%) of the scented sites[22] (Fig. 6.6).

Finally, people have known and used the beaver's castor sacs for a long time and woven stories around them. According to one fable, beavers plead for their lives and voluntarily hand over their castor sacs to the hunter, so they can live, albeit a bit diminished. Figure 6.7 reproduces a 1685 engraving illustrating this story.

Alarm Signals and Vocalizations

A nocturnal lifestyle favors communication by sound and scent. Although beavers communicate extensively by odors in their social interactions, they also use a rich repertoire of vocalizations.[23,24]

Tail Slap

The best-known alarm signal of the beaver is the tail slap. In response to any disturbance near their pond, beavers either first examine the source or if startled enough, immediately slap their tail on the water surface with a powerful stroke and dive away. The tail slap produces a sound like a flat paddle hitting a smooth water surface. Humans can hear it as far as 100 m away. Other beavers, notably the

Figure 6.7 | Beavers voluntarily handing over their castor sacs to hunter so he will spare their lives. Xylograph illustrating an old edition of Aesop's tales. Marius J., and J. Francus 1685. Castorologia. Augsburg, Germany: Koppmayer.

young, respond to an adult's tail slap by diving immediately or doing so after tail slapping in turn. Beavers on land rush to the water and submerge in response to a tail slap in the pond. They may head for the lodge.

The typical factors that stimulate tail slapping at a beaver pond are approach by a person, unfamiliar or sudden noises, and odors. For instance, during our experiments using castoreum from a strange beaver, the resident beaver tail-slapped more than usual when the strange odor was present near its lodge.[25] After diving, the beaver either reemerges soon afterward or disappears into the lodge. The dive can be noisy with splashing water or quiet with no audible sound.

Beavers discriminate tail slaps from different individuals. Tail slaps by adults, notably adult females, are more successful in eliciting flight in family members.[26] By contrast, older beavers often ignore tail slaps by juveniles that sound different, owing to their small and narrow tails. Young beavers are still learning the "social rules" of appropriately using the tail slap. Learning to discriminate tail slaps by different individuals is probably easy, as the tails differ in size and shape.

Table 6.2 shows that tail slaps by older beavers elicit more escape responses than those by younger ones. Kits induce others to flee by their tail slaps less than half as often as adults.

Table 6.2 | Tail Slaps by Adults Elicit the Most Escape Responses, and Those by Kits the Least

Tail Slaps Produced by		Escape Responses to Tail Slaps by All Age Classes	
Age Class	No. of Incidents	No.	Percentage
Adults	382	240	62.8
2- and 3-year-olds	91	40	44
Yearlings	289	134	46.4
Kits	209	62	29.7

Source: Data from reference 1.

Vocalizations

Adults Inside the lodge, beavers produce burps, whines, and gnawing and chewing sounds. Since in the beaver's life indirect sounds, produced by its movements in water, play an important role, we note here also the gargling and bubbling noises that the animals cause when leaving and entering the lodge.

Out in the pond, hissing is common, although wind may mask this vocalization. In our field studies, beavers hissed in response to strange scent marks placed by the experimenter. Leighton[27,28] heard beavers hiss in defense when they were "frightened" or "vexed," such as when they were pushed away from food or found their house closed. They also hissed toward muskrats. A captive, probably male adult, at a wildlife rehabilitator's compound, when exposed to chemical compounds from castoreum, tail slapped, hissed, and ground his teeth in response, especially to 4-ethyl phenol (our observations). We have also heard blowing in response to unpalatable food, such as witch hazel placed in the pond by the experimenter.

At night, beavers betray their wood-cutting activities by their gnawing, which is audible over considerable distances. We have never heard the call described by Grey Owl: "She [a female named Jelly Roll] often signaled me by calling, giving a clear, penetrating note capable of being heard for half a mile, and should I call her with a similar sound she seldom failed to answer me, if within earshot."[29]

Some beavers charge and hiss loudly at the same time, especially when confined in a live trap. It seems to be defensively employed to intimidate adversaries in extremely tense situations. Beavers also produce sounds assumed to express pleasure, excitement, satisfaction, and anger.[30] At this time, most of the behavioral functions of beaver vocalizations are still not fully understood.

Table 6.3 | Number of Whine Sounds Emitted and Received by Different Age Classes

Age Class	No. of Whines Emitted (E)	No. of Whines Received (R)	Ratio E/R
Adults	34	223	0.15
2-year-olds and yearlings	143	167	0.86
Kits	329	82	4.0

Source: Computed from data in reference 1.

Young Beavers Young beavers are particularly vocal. They can often be heard whining inside the lodge, even from a distance. The whine is a soft, high-pitched sound that resembles a baby cry. It appears to be related to soliciting food from other beavers. One- or even 2-year-old beavers whine not only when demanding to share food with others but also when being expelled from the lodge (see chapter on dispersal).

Hodgdon[5] recorded 508 incidents of whine calls. Younger animals whined much more frequently than older ones. Kits accounted for 64.8% of whining episodes; yearlings, 23.3%; 2- and 3-year olds, 5.3%; and adults, 6.7%. During each episode, kits repeated the whine vocalization more often (average 6.6 times) than older beavers (adults: 1.8 times). For all encounters, Hodgdon[5] also counted how often each age class whined and how often these age classes were the addressees of whining sounds. Table 6.3 summarizes his findings.

REFERENCES

1. Hodgdon, H. E. 1978. Social dynamics and behavior within an unexploited beaver (*Castor canadensis*) population [Ph.D. dissertation]. Amherst: University of Massachusetts.
2. Svendsen, G. E. 1980. Patterns of scent-mounding in a population of beaver (*Castor canadensis*). Journal of Chemical Ecology 6: 133–147.
3. Hodgdon, H. E., and R. A. Lancia. 1983. Behavior of the North American beaver, *Castor canadensis*. Acta Zoologica Fennica 174: 99–103.
4. Aleksiuk, M. 1968. Scent mound communication, territoriality, and population regulation in the beaver. Journal of Mammalogy 49: 759–762.
5. Bollinger, K. S., H. E. Hodgdon, and J. J. Kennelly. 1983. Factors affecting weight and volume of castor and anal glands of beaver (*Castor canadensis*). Acta Zoologica Fennica 174: 115–116.
6. Svendsen, G. E. 1978. Castor and anal gland of the beaver (*Castor canadensis*). Journal of Mammalogy 59: 618–620.

7. Lederer, E. 1950. Odeurs et parfums des animaux. Fortschritte der Chemie organischer Naturstoffe 6: 87–153.
8. Maurer, B., and G. Ohloff. 1976. Zur Kenntnis der stickstoffhaltigen Inhaltsstoffe von Castoreum. Helvetica Chimica Acta 59: 1169–1185.
9. Rowe, J. W., editor. 1989. Natural products of woody plants. 2 volumes. Berlin: Springer.
10. Pilleri, G. 1984. Investigations on beavers. Vol. 3. Berne, Switzerland: Brain Anatomy Institute.
11. Butler, R. G., and L. A. Butler. 1980. Towards a functional interpretation of scent marking in the beaver (*Castor canadensis*). Behavioral and Neural Biology 26: 442–454.
12. Schulte, B. A. 1998. Scent marking and responses to male castor fluid by beavers. Journal of Mammalogy 79: 191–203.
13. Schulte, B. A., D. Müller-Schwarze, and L. Sun. 1995. Using anal gland secretion to determine sex in beaver. Journal of Wildlife Management 59: 614–618.
14. Rosell, F., and L. Sun. 1999. Use of anal gland secretion to distinguish the two beaver species *Castor canadensis* and *C. fiber*. Wildlife Biology 5: 119–124.
15. Sun, L. 1996. Chemical kin recognition in the beaver (*Castor canadensis*): behavior, relatedness and information coding [Ph.D. dissertation]. Syracuse: State University of New York College of Environmental Science and Forestry.
16. Grønneberg, T. Ø., and T. Lee. 1984. Lipids of the anal gland secretion of beaver (*Castor fiber*). Chemica Scripta 24: 100–103.
17. Rosell, F., and F. Bergan. 1998. Free-ranging Eurasian beavers, *Castor fiber*, deposit anal gland secretion when scent marking. Canadian Field-Naturalist 112: 532–535.
18. Sun, L., and D. Müller-Schwarze. 1997. Sibling recognition in the beaver: a field test for phenotype-matching. Animal Behaviour 54: 492–502.
19. Sun, L. and D. Müller-Schwarze. 1998. Chemical signals in the beaver: one species, two secretions, many functions? In: Johnston, R. E., D. Müller-Schwarze, and P. W. Sorensen, editors. Advances in chemical signals in vertebrates. New York: Kluwer Academic/Plenum. p 281–288.
20. Houlihan, P. W. 1989. Scent mounding by beaver (*Castor canadensis*): functional and semiochemical aspects [M.S. thesis]. Syracuse: State University of New York College of Environmental Science and Forestry.
21. Rosell, F., and B. A. Nolet. 1997. Factors affecting scent-marking behavior in Eurasian beaver (*Castor fiber*). Journal of Chemical Ecology 23: 673–689.
22. Welsh, R. G., and D. M. Müller-Schwarze. 1989. Experimental habitat scenting inhibits colonization by beaver, *Castor canadensis.* Journal of Chemical Ecology 15: 887–893.
23. Novakowski, N. S. 1969. The influence of vocalization on the behavior of beaver, *Castor canadensis* Kuhl. American Midland Naturalist 81: 198–204.
24. Pilleri, G., M. Gihr, and C. Kraus. 1983. Vocalization and behavior in the young beaver (*Castor canadensis*). Investigation on Beavers 1: 67–80.
25. Müller-Schwarze, D., and S. Heckman. 1980. The social role of scent marking in beaver (*Castor canadensis*). Journal of Chemical Ecology 6: 81–95.
26. Hodgdon, H. E., and J. S. Larson. 1973. Some sexual differences in behavior within a colony of marked beavers (*Castor canadensis*). Animal Behaviour 21: 147–152.

27. Leighton, A. H. 1932. Notes on the beaver's individuality and mental characteristics. Journal of Mammalogy 13: 117–126.
28. Leighton, A. H. 1933. Notes on the relations of beavers to one another and to the muskrat. Journal of Mammalogy 14: 27–35.
29. Grey Owl. 1935. Pilgrims of the wild. New York: Charles Scribner's Sons.
30. Pilleri, G., M. Gihr, C. Kraus, and O. Bernath. 1975. Vocalizations and behaviour of the young beaver *Castor canadensis* (Rodentia, Castoridae). Revue Suisse de Zoologie 82: 3–26.

chapter 7

Infrastructure: Dams, Lodges, Trails, and Canals

Remarkable as the dam may well be considered, from its structure and objects, it scarcely surpasses, if it may be said to equal, these water-ways, here called canals, which are excavated through the lowlands bordering their ponds for the purpose of reaching the hard wood, and of affording a channel for its transportation to their lodges.

Lewis H. Morgan, 1868

Dams

Beavers build dams to impound water along streams. They may create one pond or several. We found as many as 18 consecutive dams in one colony, each containing a pond with a different water level. Counting tiny dams across branches of parted streams, there may even be around 40 dams on one site. On the other hand, some bank lodges along large, unchanging streams or on banks of lakes are used for a long time without any dam building.

The impoundment keeps the lodge entrances under water and permits the floating of logs and branches, diving to safety, and travel to feeding areas: the larger the circumference of the pond (or ponds), the more potential feeding range is opened up without increasing risky overland travel. Finally, beavers store branches as winter food in their ponds. From the bottom grow important food plants such as water shield (*Brassenia*) and yellow pond lilies (*Nuphar*). Beavers also defecate in the water. Thus, to the beaver the pond is highway, canal, lock (in two senses), escape route, hiding place, vegetable garden, food storage facility, refrigerator/freezer, water storage tank, bathtub, swimming pool, and water toilet.

Building materials for dams are logs of varying lengths, rocks, wads of grass, and mud. The beaver skillfully integrates these into a tightly interwoven structure. Different materials are used under different circumstances, and the shape of the dam varies with topography and water flow. One of the animal kingdom's architectural masterpieces, a beaver dam can cross an entire valley bottom or successfully withstand the pressures of a steep mountain stream.

Dams are as low as 20 cm or as high as 3 m; they can measure 1 foot (0.3 m) or several hundred meters in length. A dam can collect mere seepage into a coherent, though shallow, and often ephemeral body of water; can be built on top of a man-made concrete outlet of a lake; or can stretch across roaring mountain streams. The optimal dam, however, contains slowly moving water with a stream gradient of 1% or lower.

To understand the design features and functioning of a dam, it is best to observe the beaver starting a new dam. We have been fortunate to witness this several times. In our observations, the technique applied depends on the water flow: In standing shallow water, beavers raise the level gradually by depositing mud, leaves and small twigs. By contrast, a large stream requires a solid base for the new dam. As Plate 14 shows, sticks about 2 m long and 5 cm in diameter are propped against the banks and point upstream at an angle of about 30 degrees. The beavers weigh down these poles with heavy stones (Plate 15). Grass is stuffed between the rocks. The animals float poles through the pond, clamber on the dam, and ram these sticks into the existing dam. They align the poles with the direction of the water flow. Seen from the downstream side, the dam presents a huge structure of parallel sticks extending from the crest of the dam down to the flowing water. Over time, as new material is added and the dam settles, a tight, compacted slope develops on the pond side of the dam. This impressive structure is revealed when a pond drains (Fig. 7.1). Where little water flows, the animals dredge sediment up from the pond and lay down a generous layer of mud on the dam (Plate 16) and thus retain more water. During droughts we observed the

Figure 7.1 | The massive inner wall of a dam is revealed when a pond drains.

Figure 7.2 | Bank holes in a drained pond.

beavers raising the dam by adding mud. Presumably, their response is adaptive: they "hog" scarce water.

The sound of running water appears to be a strong stimulus to building or repairing a dam. For this reason, flow control devices should produce no or little noise, or else the beavers will try to plug them up. Trappers exploit the beaver's response to rushing water by cutting a notch into the dam and placing their trap there. The animals will invariably try to repair the dam at that spot. Beavers build dams in late summer and fall, but they repair dams whenever necessary. For instance, Townsend[1] observed most dam building around the first of September.

Beaver dams are superior to man-made concrete dams in retaining water and releasing it slowly. Especially leaky beaver dams can serve as ideal water flow regulators (D. Hey; personal communication, 2000). Leaky or not, dams in steep terrain sometimes reach the impressive height of 3 m (Plate 17).

Lodges

The lodge is the focal point of a beaver colony. It is the principal shelter for the family, providing protection from cold, heat, and predators. Here the beavers rest, sleep, mate, bear the young, and rear them. When alarmed, beavers dive into the lodge, and they also bring food into the lodge and consume it there.

First beavers dig and use *bank holes,* tunnels in steep slopes, with the entrances under water (Fig. 7.2). They can elaborate such a burrow into a bank lodge (Plate

Figure 7.3 | Lodge, freshly mudded in fall. A food cache on the left.

18) by piling sticks over and around the entrance and on a hole that may form in the "roof" of the bank hole. The ultimate structure, however, is the freestanding lodge, entirely surrounded by water, and hence safe from land predators. All three types of shelters may be found at a given beaver site.

In a lodge the beavers rest on a platform above water. The structure has several entrances under water. Since the resting platform inside the lodge must be above the water level, with the entrances submerged, a freestanding lodge is established in shallow water and often constructed over a "feeding platform," a place surrounded by water where the beavers sit on their haunches and eat bark from logs and sticks.

Building materials are sturdy logs and mud. Especially in fall, the lodge is mudded up and sealed, except for an air vent at the top (Fig. 7.3). In spring, after rains have washed off part of the mud, the lodge is repaired. Although mostly peeled logs are used, some logs still have their bark on.

The size of a lodge can vary from a tiny, sloppy "trial lodge" of a young first-time settler (see Plate 29) to a massive family fortress that measures 6 m in diameter or more at the water line (and much more at its true base) and over 2 m high.

Lodge building has been little observed, since it takes place mostly during the night. Speed is the hallmark of the beaver's work; a substantial lodge for the winter can be built in only 2 nights.

Temperatures

The lodge insulates the beavers extremely well from outside temperatures. In the summer, beavers avoid open-water lodges and prefer the cooler bank lodges instead. From April to October, the inside of the lodge is about 2°C cooler than the surrounding air.[2] When the outside temperature drops to –21° to –6.8°C in winter (January and February), the air inside remains at 0.8°–1.6°C. These temperature ranges are average daily minimum and maximum temperatures, respectively. The lodge temperatures resemble those of the surrounding water, which is about 0.5°C in winter. We also see from these data that the temperature in the lodge remains very stable. It varies by only 0.8°C in 24 hours, while the outside temperature fluctuates by 14.2°C. The inside temperature fluctuates slightly when the lodge lacks snow cover and when beavers leave and enter the lodge.[3] The thick layer of mud that beavers apply to the lodge mitigates the cooling effect of the cold winter winds. This is helped further by snow cover. In winter, thin-walled lodges are 7°–8°C cooler than thicker-walled ones. In summer, a thinner wall provides the advantage of better ventilation.[2] One wonders how lodges without an outer cover of mud fare during the winter. Along northern lakes with little sediment, beavers do not easily find mud and therefore do not plaster their lodge with mud at any time of the year (Plate 19). This circumstance seems to require sociality: in terms of thermoregulation, a single beaver may be at a serious disadvantage during winter. Beavers also need their lodge for warming up after winter excursions in the water. While they forage under the ice, body temperature can drop below 34°C. Back in the lodge, the beaver returns to normal body temperature within 60 minutes.[4]

Ventilation

The lodge is well ventilated. The levels of carbon dioxide and oxygen inside the lodge do not vary over the seasons.[5] Experimentally added carbon dioxide is cleared away within 60 minutes.[5] During summer, the lodge walls serve in ventilation. In winter, the vent at the top of the lodge becomes more important, as the walls are mudded, frozen, and covered with snow.

Species Differences

Construction behavior is more developed in the North American beaver. In some places in Russia, both the Eurasian and North American beavers live side by side. Under the same ecological conditions, the Eurasian beaver is less likely to build dams and lodges (Table 7.1).[6]

Modification under Changing Water Levels

Beavers have to adjust the structure of their lodges to varying water levels. If the water level drops, they extend the lodge into the pond or lake so that the entrances remain underwater. Rising water levels present more of a challenge. The

Table 7.1 | Colonies of Eurasian and North American Beavers with Dams and Lodges in Different Populations in Northwestern Russia

Species	Percentage with Dams	Percentage with Lodges
Eurasian beaver	18.0–53.6	12.5–47.1
North American beaver	74.8–100	54.5–66.9

Source: Reference 6.

most extreme example was the large-scale flooding of beaver colonies during the establishment of the vast hydroelectric project "La Grande Complex" undertaken by the James Bay Energy Corporation in Quebec, Canada. Nearly 9700 km^2 was flooded in the 1970s and 1980s. Radio-tagged beavers remained in their home ranges during the flooding. Every time their lodge was flooded, they built a new lodge. In early winter, building a new lodge became difficult. The beavers solved their problem by adding an "upper floor" to their lodge.[7] Some similar observation might have given rise to the myth of multifloor beaver lodges, as depicted in the historical engraving of Figure 7.4.

Figure 7.4 | Multilevel beaver lodge, as imagined in a 1750 engraving of Indians hunting beavers, by an unknown artist. (Reprinted with permission from Amon Carter Museum, Fort Worth, Texas.)

Figure 7.5 | Canal between lodge and feeding grounds (in foreground). Acadia National Park, Maine.

Trails

Wherever beavers repeatedly forage away from water, they wear down vegetation and create trails (See Plate 6), especially by dragging tree limbs to the water. The trails can be short, merely pathways from the water to solid land. At Allegany State Park, "regular" feeding trails measured from 15 to 18 m. Colonies with depleted tree resources around their pond went farther afield: six trails to upland aspen groves averaged 129 m (range: 65.3–201.0 m).[8]

Canals

On more or less level ground, well-worn trails will fill with water. Beavers further improve these incipient canals: they dredge up mud and deposit it at the banks of these channels. The canals ease the transport of logs from the foraging sites to the lodge and the dam (Fig. 7.5).

A beaver family dredges out channels at the bottom of their pond. These come in handy in times of drought when ponds dry out. The channels then provide waterways in an otherwise useless mudflat.

REFERENCES

1. Townsend, J. E. 1953. Beaver ecology in western Montana with special reference to movements. Journal of Mammalogy 34: 459–479.
2. Buech, R. R., D. J. Rugg, and N. L. Miller. 1989. Temperature in beaver lodges and bank dens in a near-boreal environment. Canadian Journal of Zoology 67: 1061–1066.
3. Stephenson, A. B. 1969. Temperatures within a beaver lodge in winter. Journal of Mammalogy 50: 134–136.
4. MacArthur, R. A., and A. P. Dyck. 1990. Aquatic thermoregulation of captive and free-ranging beavers (*Castor canadensis*). Canadian Journal of Zoology 68: 2409–2416.
5. Dyck, A. P., and R. A. MacArthur. 1993. Seasonal variation in the microclimate and gas composition of beaver lodges in a boreal environment. Journal of Mammalogy 74: 180–188.
6. Danilov, P. I., and V. Ya. Kanshiev. 1983. The state of populations and ecological characteristics of European (*Castor fiber* L.) and Canadian (*Castor canadensis* Kuhl.) beavers in the northwestern USSR. Acta Zoologica Fennica 174: 95–97.
7. Courcelles, R., and R. Nault. 1983. Beaver programs in the James Bay area, Quebec, Canada. Acta Zoologica Fennica 174: 129–131.
8. Müller-Schwarze, D., and B. A. Schulte. 1999. Behavioral and ecological characteristics of a "climax" population of beaver (*Castor canadensis*). In: P. E. Busher and R. M. Dzięciołowski, editors. Beaver protection, management, and utilization in Europe and North America. New York: Kluwer Academic/Plenum. p 161–177.

chapter **8**

Beaver Time

Beaver do not surrender themselves to the confines of a house and pond until cold solidly covers the pond with a roof of ice. The time of this is commonly about the first of December . . . an unusually early winter or even a heavy snow may so hamper them that, despite greatest effort, the ice puts a time lock upon the pond and closes them in for the winter without sufficient supplies.

Enos A. Mills, 1913

When a beaver is active and when it rests depends largely on three factors: exposure to the natural light cycle, air temperature, and season. Beavers stay in their lodge during the daytime, from about 0800 to 2000 hours in the summer in northern latitudes. During the first part of the night they feed, and during the second half they construct dams and lodges.

Most surprising is the beaver's ability to decouple its daily activity rhythm from the natural 24-hour light cycle. Western colonies in Alberta, Canada,[1] as well as eastern ones in Quebec,[2] have shown a free-running cycle of activity that deviates considerably from 24 hours. Such free-running cycles are usually observed only under artificially constant light conditions (continuous light or darkness) in the laboratory. The beaver, when confined to its lodge at times of heavy ice cover on their pond (Plate 20), may have an activity cycle of about 27 hours in Alberta and one that may range from 26.25 to 28.0 hours in Quebec (Fig. 8.1). In the studies just cited, the activity of the family inside the lodge was recorded with the aid of a microphone.

Lancia[3] placed radio-collars on beavers and was thus able to record the activity of individuals separately. In winter, only the female lengthened her activity cycle to 26.58 hours, while the male remained on a 24-hour cycle. Lancia suggested that the female's activity rhythm is related to the onset of her estrous cycle.

The beaver is not unique in showing a free-running activity period that is much longer than 24 hours. Other such animals include the minnow (*Phoxinus phoxinus*),[4] burbot (*Lota lota*),[5] trout (*Salmo trutta*),[6] and yellow-necked field mice (*Apodemus flavicollis*).[7] The free-running rhythms of these free-living species range from 24.2 hours for the field mouse to 29 hours for trout. These cycles

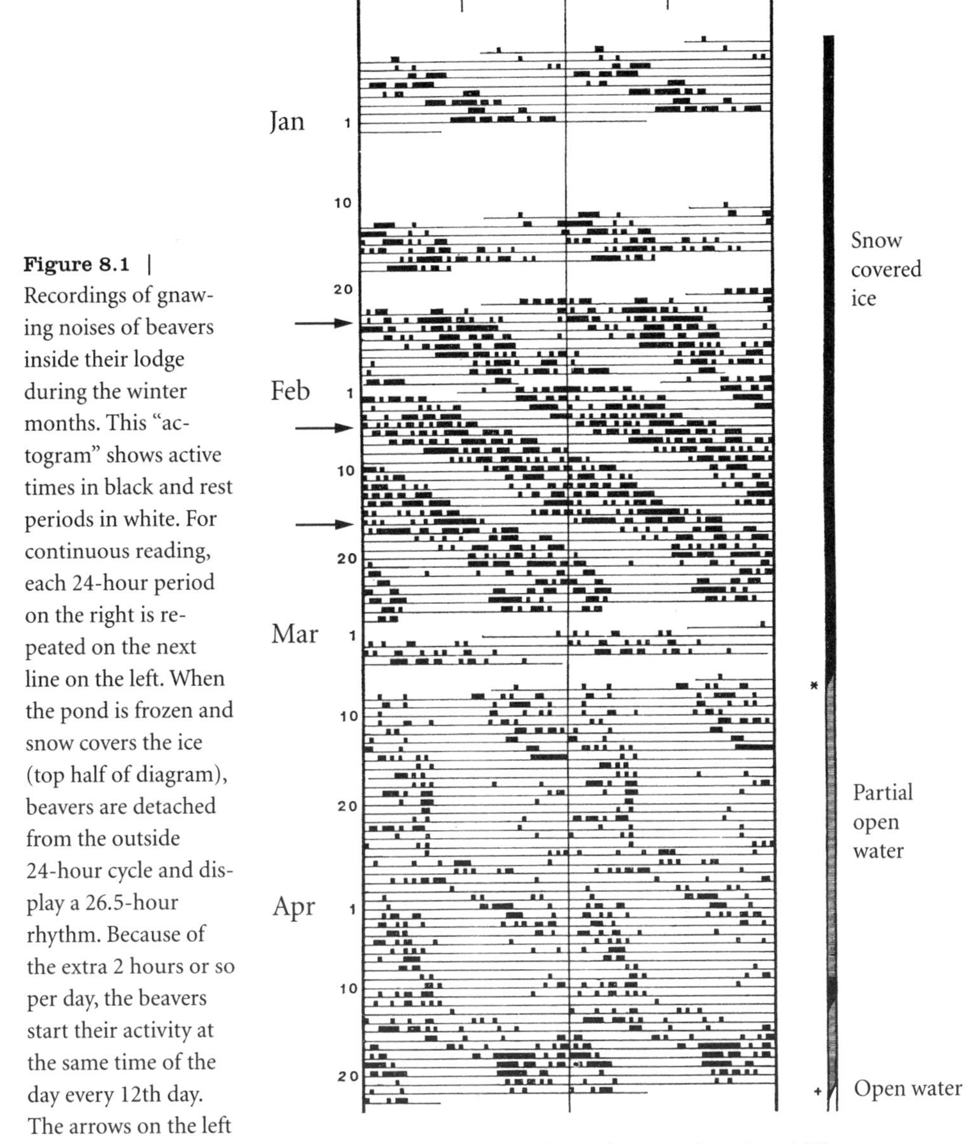

Figure 8.1 | Recordings of gnawing noises of beavers inside their lodge during the winter months. This "actogram" shows active times in black and rest periods in white. For continuous reading, each 24-hour period on the right is repeated on the next line on the left. When the pond is frozen and snow covers the ice (top half of diagram), beavers are detached from the outside 24-hour cycle and display a 26.5-hour rhythm. Because of the extra 2 hours or so per day, the beavers start their activity at the same time of the day every 12th day. The arrows on the left mark the days when this happens: The activity starts at the 24-hour mark in the middle of the graph. In March and April (lower half of graph) the water surface is partially open; a 24-hour rhythm occurs from March 12–24. After a period with a cycle of 26.5 hours, the 24-hour cycle resumes around April 18 when the pond is completely open. The asterisk indicates when the beavers leave the lodge for the first time; the cross, when the dam breaks, and the beavers leave the site. (Modified from Journal of Comparative Physiology, Annual cycle of patterns of activity rhythms in beaver colonies (*Castor canadensis*), C. L. Potvin and J. Bovet, vol. 98, p. 254, Fig. 6, 1978, © Springer-Verlag.)

are considerably longer than those under constant conditions in the laboratory, which range from 23 to 26 hours.

In summer, beavers are synchronized with the 24-hour day. Even in winter, when during a thaw open water exposes them to the light regime of the external world, beavers temporarily revert back to a regular 24-hour cycle.[2] During a regular 24-hour cycle, beavers divide their time between activity and rest about evenly.[8] A family becomes synchronized with the environment but not necessarily with other colonies. At this time, we lack population-wide studies of activity rhythms under varying ice conditions.

In winter, above-ice activity ceases at certain critical low temperatures. This limit varies between geographical areas. In Sweden beavers were not active above the ice when the temperature was below –6°C.[9] In the northeastern United States (Massachusetts), this critical threshold was –10°C,[3] and in the former Soviet Union, –18°C.[10]

REFERENCES

1. Bovet, J., and E. F. Oertli. 1974. Free-running circadian activity rhythms in free-living beaver (*Castor canadensis*). Journal of Comparative Physiology 92: 1–10.
2. Potvin, C. L., and J. Bovet. 1975. Annual cycle of patterns of activity rhythms in beaver colonies (*Castor canadensis*). Journal of Comparative Physiology 98: 243–246.
3. Lancia, R. A. 1979. Year-long activity patterns of radio-marked beaver (*Castor canadensis*) [Ph.D. dissertation]. Amherst: University of Massachusetts.
4. Mueller, K. 1968. Freilaufende circadiane Periodik von Elritzen am Polarkreis. Naturwissenschaften 55: 140.
5. Mueller, K. 1969. Tagesperiodik von Wasserorganismen am Polarkreis. Umschau 1969: 18–19.
6. Mueller, K. 1969. Jahreszeitliche Wechsel der 24h-Periodik bei der Bachforelle (*Salmo trutta* L) am Polarkreis. Oikos 20: 166–170.
7. Erkinaro, E. 1969. Der Verlauf desynchronisierter, circadianer. Periodik einer Waldmaus (*Apodemus flavicollis*) in Nordfinnland. Zeitschrift für Vergleichende Physiologie 64: 407–410.
8. Lancia, R. A., W. E. Dodge, and J. S. Larson. 1980. Summer activity patterns of radio marked beaver, *Castor canadensis*. In: C. J. Amlaner, Jr. and D. W. MacDonald, editors. A handbook of biotelemetry and radio tracking. Oxford: Pergamon. p 711–715.
9. Wilsson, L. 1971. Observations and experiments on the ethology of the European beaver (*Castor fiber* L.). Viltrevy 8: 115–266.
10. Semyonoff, B. T. 1957. Beaver biology in winter in Archangel Province. In: Translations of Russian game reports. Volume 1, Beaver 1951–1955. Translated by J. M. MacLennan. Ottawa: Canadian Wildlife Service. p 71–92.

Food Selection

> Their food chiefly consists of a large root, something resembling a cabbage-stalk, which grows at the bottom of the lakes and rivers. They eat also the bark of trees, particularly that of the poplar, birch, and willow; but the ice preventing them from getting to the land in Winter, they have not any barks to feed upon during that season, except that of such sticks as they cut down in Summer, and throw into the water opposite the doors of their houses; . . . the roots above mentioned constitute a chief part of their food during the Winter. In summer, they vary their diet, by eating various kinds of herbage, and such berries as grow near their haunts during that season.
>
> *Samuel Hearne, 1795*

Range of Plants Eaten

Beavers fell trees. They shock us by cutting down a specimen tree in our backyard, or make national news by destroying cherry trees at the Tidal Basin in Washington, D.C. At the very least, many of us have seen stumps left by beavers or pruned willows along a stream. Indeed, many professional studies of food habits have focused on stumps, severed tree trunks or limbs on the ground or in the water, and food caches, which are piles of branches from various trees that beavers conveniently place near their lodge for the winter. This woody material is literally hard evidence of the beavers' foraging and feasting.

Direct observation of the beavers' feeding behavior teaches us that they consume much nonwoody vegetation that leaves little trace. Besides trees, beavers eat grasses and forbs on land and aquatic vegetation in the pond or at the lake bottom. Of these three categories of plant food, beavers in southeastern Ohio spent 60%–90% of their feeding time on tree bark in March/April and October/November. In May they abruptly switched to grasses, and later in summer to aquatic plants. During the summer months, they fed on nonwoody vegetation 90% or more of the time.[1] From March to November (including summer), one Ohio colony spent only 40.2% of feeding time on woody vegetation, another one even only 31.4%, and the remainder was spent eating grasses and forbs.[1] These beavers may have chewed additional bark while inside their lodge. In Pennsylvania, the ratio between woody and nonwoody foods was 25:6.2 kg/month in winter, but 2:30 kg in summer.[2] Therefore, to appreciate the needs and impact of beavers, it is imperative to consider all types of food plants. Indeed, where available, beavers prefer herbaceous vegetation, such as water lily rhizomes, to woody vegetation in all seasons.[3]

In agricultural areas beavers consume corn, and they readily accept other foods such as apples when offered by well-meaning people.

Where Beavers Harvest Trees

Beavers are "central place foragers": from the lodge where they live, they venture out in all directions to cut plants. They forage at greater distances upstream from their main lodge than downstream.[4] Obviously, it is easier to float logs downstream. On the slopes near their site, beavers forage the less the greater the distance. This way they minimize not only energy expenditure but also risk of predation. However, once areas close to the pond become depleted of preferred tree species, beavers have to go farther afield.[5]

Beavers clear-cut at distances of up to 300–500 feet (92–152 m) from their pond. This distance affords tolerable risk and energy expenditure. For a stand of aspen they may go as far as 200 m from their main pond, and up steep slopes at that.[6] Generally speaking, beavers forage more selectively at increasing distances from their "central place."[7] In one study in Michigan, beavers used 21,313 m^2 of terrestrial area around the pond for cutting woody vegetation.[8]

Tree Preferences

North American beavers prefer aspen and other species of the genus *Populus*. Indeed, a look at the distribution maps of beavers and quaking aspen (*P. tremuloides*) confirms that the ranges of the two species are almost identical. Beavers are also fond of willows.[9] Where aspen have been depleted locally near beaver sites, willow becomes the main food. This is the case in Rocky Mountain National Park and at some sites at Allegany State Park in New York, to name two examples. The first beaver family to establish itself at Allegany State Park in 1937 found plenty of choice among various tree species. They cut 199 quaking and bigtooth aspen trees (71% of all trees harvested) but only 3 small willows (1.1%).[10] Similarly, beavers on a willow-dominated site (36.6% of all trees) in Ohio preferred aspen and alder. Willow constituted only 5.9% of their diet.[11] At their northern distribution limit, such as in Alaska and Labrador, beavers subsist on willow, for lack of other tree species.

Beavers need a mixed diet. Kept on a strict diet of quaking and bigtooth aspen for several weeks, American beavers lost weight at the rate of 0.1%/day. Fed only red maple, paper birch, or alder, the animals lost even more weight (0.3%–0.6%

of body mass/day). One beaver actually died on such an extremely restricted, one-species diet.[12]

What tree species beavers choose depends on the locally available mix. As "choosy generalists," they are highly flexible at any one locale and over their entire geographical range. However, just as aspen and willow usually rank at the top, their last choices predictably are conifers such as spruce, fir, and pine, although on occasion these are eaten to some extent too. Among the deciduous trees, red maple (*Acer rubrum*) scores very low, right along with the conifers.[13,14] Many of the other forest trees, such as beech, ash, maples, hornbeam, and cherries, fall somewhere in the middle. The preference order varies, depending on local abundance of the various trees. For instance, on Isle Royale in Lake Superior, beavers preferred sugar maple and yellow birch, while they tended to avoid paper birch, baked hazelnut, honeysuckle, balsam fir, northern white cedar (arborvitae), spruce, and elder.[8]

Where introduced into new areas with tree species entirely "exotic" to them, beavers have no problem living on food that does not occur in their natural range. Transplants to Tierra del Fuego use several species of southern beech (genus *Nothofagus*) as food and building material (see Plate 44). Thriving on this, they have colonized many rivers and today live in a high-density population.

Different beaver families in the same population may prefer different food plants. For example, of two colonies on Isle Royale that were offered food experimentally, only one accepted white clover readily, while only the other ate yellow pond lily stems and leaves. Each chose the plant that grew naturally at their site and ignored the species that was absent.[15]

Food Choice and Feeding Behavior

First, beavers sample trees. Careful observation near beaver ponds reveals saplings and trees beavers have bitten into but left alone (Fig. 9.1). They appear to assess nutritional value. In a study in Massachusetts, beavers switched from birch to oak and witch hazel in two subsequent fall seasons. During the second season they sampled many birch trees without utilizing them further.[16]

How do beavers choose one species over another? In an ingenious experiment, Doucet and coauthors combined aspen stems and "canopies" of red maple and then red maple stems and aspen canopies.[17] After this switching of canopies, beavers judged these "hybrid plants" by their stems: they cut more aspen stems with maple canopies than the other way around. Covering the stems with paper bags did not alter their response. This shows that odor, rather than visual cues, is important in beaver food selection.

Distance from the water also plays an important role: the greater the distance from the pond, the more selective the beavers are in terms of species chosen (they go far for aspen), and the smaller the diameter of the trees the beavers cut.[18]

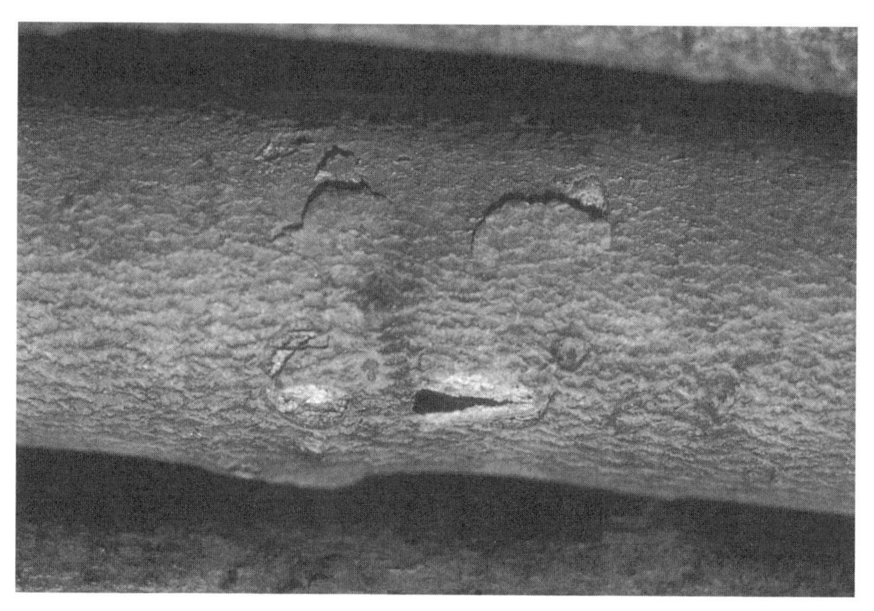

Figure 9.1 | Red maple sampled by beavers (tooth marks).

Size of trees and palatability are interrelated: beavers cut only smaller specimens of less preferred trees, while they take all size classes of the more preferred trees. This has been interpreted as the result of the smaller nonpreferred trees ("best" sizes of nutritionally poorer species) being nutritionally equivalent to the "worst" sizes of the preferred species.[19]

Beavers often eat bark of thicker trees directly on site, without felling the tree itself. Of those they cut down, they gnaw thicker branches into pieces that can be easily hauled or floated to the water area near the lodge (Fig. 9.2). They chew the bark off while holding the stick or log in their front paws (Plate 21). Where beavers continue to consume the bark of twigs and logs in shallow water over several days, a "feeding bed" results (Fig. 9.3). The sticks in various stages of being peeled provide evidence of what species are preferred at that site and time. Also, very thin sticks as leftovers indicate that kits might be present.

How long does it take a beaver to cut down a tree? For all woody plants, the overall average time per plant was 1.24 minutes, according to one study.[8] Trees up to 15 cm in diameter can be felled in less than 50 minutes. But the time needed for larger trees increases exponentially: trunks 25 cm and larger require over 250 minutes.[8] After all this effort, many trees become hung up in neighboring trees. How often this happens depends on the density of the stand of trees. In Europe, 12.5%–15.0% of beaver-cut trees end up not being available for this reason.

Figure 9.2 | Aspen logs, cut into typical length for transport.

Figure 9.3 | A "feeding bed" in a beaver pond.

Figure 9.4 | Beavers have dug up the ground near their pond to feed on bracken rhizomes. Pond in background.

Seasonal Changes of Food Preferences

In addition to the alternation between trees in winter and grasses, forbs, and aquatics in summer, beavers prefer different tree species during different seasons. Conifers such as pine are accepted primarily in late winter and spring[1,3] (our own observations) (Plate 22). Presumably, lack of other vegetation, combined with renewed sugar flow in the pines, renders them attractive for a short period of time. In fact, we found a colony that used Scots pine in April. In a choice experiment, this family preferred Scots pine to aspen! Also, beavers might need compounds from pine bark for forming castoreum, which they use for scent marking in spring.[3] In early spring our beavers at Allegany State Park, and also in the Adirondacks, dig up the rhizomes of bracken fern at the edges of their ponds (Fig. 9.4). Beavers intensify tree cutting in the fall. In Massachusetts, the number of trees cut by one colony increased drastically from mid-October to late November.[18]

Nutrients

Proteins, carbohydrates, and fats form the foundation of an adequate diet. While animals choose vegetation of high quality, many plant species contain little of one or more of the essential nutrients. This forces herbivores to subsist on a varied diet. The beaver exemplifies such a generalist feeder.

Nutritionally, tree species differ. For instance, the often-rejected red maple has less energy content[20] (see Digestion and Digestibility).

In addition to these basic nutrients, vitamins and trace elements are important. Deficiencies can lead to disease and death. The famous captive beaver "Nicky" at Beaversprite in New York died of goiter in September 1987. Three other beavers had died of goiter before. It is well known that captive wildlife such as deer and felines develop goiter, an iodine deficiency. In a study in Michigan, the iodine content of deer foods ranged from a minimum of 0.008 part per million (ppm) in fruits, nuts, and grains to 3.1 ppm in aquatic plants.[21] Now we can see how important those aquatics must be to beavers. Water plants also can provide beavers and deer with sodium in areas with sodium-deficient terrestrial vegetation. However, beavers in Algonquin Provincial Park in Canada[12] and on Isle Royale National Park in the United States[14] do not seem to suffer a sodium constraint.

Secondary Plant Compounds

In addition to nutrients and trace compounds and elements, plant secondary compounds also affect the beavers' feeding, often in a negative way. Phenolic-laden trees such as red maple and witch hazel are avoided (Plate 23). Especially young trees can be chemically defended.

In a comparative study, beavers selected smaller-diameter aspen at Little Valley, Nevada, but preferred larger diameters at Sagehen Creek, California (Fig. 9.5). The two sites differed in that beavers had colonized Little Valley only recently, whereas beavers had cut aspen at Sagehen Creek for over 20 years (as described in Hall's study[9]). Sagehen Creek experienced considerable juvenile regrowth of as-

Figure 9.5 | Aspen cut at Sagehen Creek, California.

pen, which is characterized by high levels of a phenolic compound.[22] Feeding experiments showed that beavers avoid the root sprouts of aspen that have juvenile characteristics, including large leaves, and no lateral branching. A phenolic factor with a molecular weight of 426 occurs at high levels in juvenile sprouts, but it is not one of the known compounds salicin, salicortin, tremuloidin, and tremulacin. These four compounds occur at approximately the same levels in adult-form and juvenile-form sprouts.[23] The phenolic factor may be responsible for the discrimination of these two growth types by beavers.

Beavers cope with one class of phenolics, the tannins, in an evolutionarily ingenious way. As generalist feeders, they encounter many different phenolics. However, they employ physiological coping mechanisms for only the most characteristic tannins of their diet. These are *condensed* (as opposed to *hydrolyzable*) tannins. Condensed tannins, in turn, can be linear molecules or have a branched structure. Beavers prefer tree species with linear, and not branched, condensed tannins. Correlated with this, beaver saliva contains proteins that bind only to the linear and not to the branched condensed tannins and hydrolyzable tannins.[24]

Herbivores in general may feed on willow because this way they avoid the chemical defenses, mostly resins, of other deciduous trees such as birches and alder.[25]

Food Conditioning

Beavers leave twigs and sections of tree limbs in the water before they consume the bark. In the winter food caches, branches also sit in the water for weeks or months. The beaver is in an excellent position to utilize streams and its ponds for leaching out undesirable compounds from the bark before eating it. The first experiments to test whether beavers process their food in this way suggested that they do, but we are far from knowing for sure. In Upstate New York, beavers tend to leave sticks of the less palatable red maple and witch hazel in the water for 1–3 days before they consume the bark.[26]

Digestion and Digestibility

Aspen passes through the digestive tract of beavers in about 10–20 (mostly 16–18) hours, while red maple takes 30–50, even as long 84 hours.[12,20] The faster the food passes through the beaver's system, the higher the rate of nutrient absorption. This explains why beavers accept red maple very reluctantly, and only when they have to. Because of the limitations on how much of each food can pass through the gut each day, Fryxell and Doucet[20] calculated that the beavers could never meet their energy requirement were they to eat red maple exclusively.

Since some plants are more digestible than others to the beaver, we can determine the digestibility by measuring how much the beaver eats and comparing this with the fiber or nitrogen content of the droppings. The difference is the digested portion of the food. The ratio of the amount of matter or energy digested to what is eaten is termed *apparent digestibility.*

However, some material excreted in the feces does not stem from food but rather from the animal's body itself. If such "endogenous losses" are taken into account, "true digestibility" can be determined. This measure is more useful than apparent digestibility.[27] True digestibility can be determined by plotting the ingested daily amounts of, say protein for each animal on the abscissa, versus the digested amount per day on the ordinate of a graph. Such a graph shows that red maple, for instance, is eaten in smaller amounts than aspen or alder, and for the amount eaten, less of the protein is digested.[27]

These studies show that several different factors affect how beavers select trees. Size, distance, species, nutritional value, and palatability all play a role.

Beavers do not seem to possess cellulases, the enzymes to digest cellulose. And yet they are able to digest about 30% of the cellulose they ingest, most likely by the action of microorganisms.[28] It is not clear whether the cardiac gland along the lesser curvature of the stomach plays a role in cellulose digestion. Although they consume more woody material than most other mammals, beavers have no special adaptations for cellulose digestion beyond what we find in ruminants.[28]

Caecal microbes may turn cellulose into nutrients that the beavers take up from their "cloaca" and eat a second time. This reingestion of "coecotrophe" is neither rumination nor coprophagy, as the caecal material is blackish and soft and therefore quite different from regular feces. During this second passage, the small intestine will absorb the recovered nutrients. The smaller particles of this reingested coecotrophe travel faster on their second passage through the gastrointestinal tract. When labeled with magnetic, fluorescent microtaggant, 55% of these smaller particles passed after 10–14 hours, and 88% passed after 40 hours. By comparison, only 6% of eaten bark was digested after 11 hours (first defecation after a meal), but 88% was digested by the second defecation on the second day. In summary, passage time for 100% of both original food and reingested material is about 60 hours. But the mean passage time for the microtaggant was only 36 hours. This is a turnover rate of 70% of eaten material in 24 hours for a beaver on an aspen bark diet.[29] Compare this with some extremes in birds and mammals: seeds can pass through songbirds in only 12 minutes but will take up to 60 days in a horse.

Estimates of Food Supply

Wildlife managers and property owners often want to know whether an area harbors an adequate food base for beavers. For proper estimates we need to know not only the food preferences of the beavers but also how flexible beavers are in terms of making do with less preferred species. To derive an estimate of available beaver forage from a count of aspen or willow specimens, one has to consider how much of the total biomass is edible. For instance, 93.6% of the stem weight of first-year shoots (2.5 mm in diameter or less) of coyote willow (*Salix exigua*) is beaver food. By contrast, the edible bark of the largest investigated willow stems (40–60 mm in diameter) comprised only 12.2% of their total biomass.[30]

Edible biomass, consisting of leaves, twigs, and bark, has been calculated for trembling and bigtooth aspen of different sizes. Trees with a breast-high diameter (bhd) of 2.5 cm (1 inch) provide 1.3 kg (2.85 lb) of beaver food on average; those with a bhd of 7.5 cm (5 inches), 21.3 kg (46.9 lb); and trees with a bhd of 25 cm (10 inches), 101.2 kg (223 lb).[31]

Food Caches

In northern areas, beavers store food for the winter in a cache or "raft" (see Plate 4). In fall (October/November) they drag branches and saplings to their pond and pile them up in the water close to the lodge. This way the food can be reached easily from the lodge and eaten or removed under the ice. The time of cache construction varies with latitude. A beaver population on the Mackenzie Delta (latitude 69° north) in the Northwest Territories, Canada, close to the northern limit of the species' distribution, collected their food caches in late August and early September. They stored mostly willow (76% of the cache) but also poplar (*Populus balsaminifera;* 14%) and alder (*Alnus crispa;* 10%). Willow was preferred, while alder was taken much less than its abundance would suggest.[32]

In another study in Wood Buffalo National Park (latitude 58° north) in Alberta, Canada, beavers also stored mostly willow, specifically sand-bar willow (*Salix interior*). Here the beavers are icebound during the winter for about 164 days, almost half a year. The butt diameter of the stored branches rarely exceeded 5 cm. The available amount of food in five caches was estimated to range from 39.5 to 56.0 kg (87–143 lb). Willow, alder, aspen, and red osier were analyzed for protein, fat, fiber, and nitrogen-free contents, and the digestible energy in each cache was calculated. The five food caches contained from 30,000 to 70,000 kcal. Compared to the needs of the beaver in winter, this does not suffice. The older age groups run a deficit of 57,000 to 150,000 kcal/colony during the 150 days they are icebound.[33]

In the northern interior of British Columbia (lat # 54°N to 55° 30'N) white spruce and subalpine fir form the forest, with trembling aspen stands interspersed. Along the waterways willow and alder are common. In fall, before consumption, 83% of the food caches contained aspen; 50%, willow; and 44%, alder. Aspen and willow were placed throughout the cache, while alder occurred only in the raft, that is, on top of the cache. Low-preference or nonfood species are also often found in the raft covering the cache. These include white birch (*Betula papyrifera*), subalpine fir (*Abies lasiocarpa*), and white spruce (*Picea glauca*). After use of the stored food (in spring), 78% of the caches contained unbrowsed alder;

28%, aspen; and 18%, willow. Thus, beavers first select aspen from the cache. The uneaten alder was later used as construction material. These findings show that the stumps of trees cut by beavers or the composition of the food cache do not necessarily indicate what the beavers actually eat.[34]

Farther to the south, in west central Massachusetts, the numbers of branches in the food cache reflected the relative abundance of the various tree species at the two investigated colonies. At one colony, sugar maple (*Acer saccharinum*) and alder (*Alnus rugosa*) comprised 61.2% of the total available trees and 79% of the branches in the food caches. Black birch (*Betula lenta*) and alder were preferred, and ironwood was avoided. At the other colony, sugar and red maple made up 68.5% of the trees present and 91.8% of all branches stored. Sugar maple was significantly preferred, and white pine (*Pinus strobus*) selected against. Shrubs (winterberry, *Ilex verticillata*, and blueberry, *Vaccinium* spp.) were also collected. They made up 52% of all branches.[35]

In a cafeteria-style experiment, Busher[35] offered five tree and one shrub species to beavers from mid-October to early December while the beavers were constructing food caches. He followed the fate of the branches that beavers removed from the experimental piles on the bank of the pond. The beavers stored more than 50% of the branches of yellow birch, white pine, red maple, and the shrub witch hazel. They ate immediately more than 50% of the red oak branches and divided the black cherry branches about equally between storage and immediate consumption. Early in the season (mid-October), beavers were more selective in caching different tree species than later (late November), when they cached almost everything offered. Several tree species were not incorporated into the cache when beavers foraged for themselves, but were stored when offered experimentally. These were black cherry, yellow birch, and white pine. Busher[36] pointed out that diversity of food plants in the cache may be adaptive.

Beavers venture farther from the pond from September to November when they construct food caches. In Pennsylvania, they traveled up to an unusual 600 m from the pond to get food for storage.[37]

Captive beavers construct food piles with whatever is available to them. One particularly interesting case was about two young beavers that were raised without an opportunity to learn from older beavers. They severed the branches from supplied aspen and cut the remaining stems into pieces about 1 m long. These pieces floated. The beavers lashed them together with smaller branches. The animals cut up other branches, dove under the raft, and left them beneath the raft. They did this for several days and nights until a compact pile had accumulated. Eventually, the raft became waterlogged and sank below the surface, holding the food pile in place where the ice would entangle it under normal circumstances. However, the water in their concrete basin did not freeze during the winter because a steam pipe ran along one wall, unknown to the experimenters. The fol-

Figure 9.6 | Array of experimental sticks nailed to a long pole before contact with beavers.

lowing fall, the beavers did not build a food cache, nor did they ever in any of the 6 subsequent years.[38]

Cafeteria-Style Food Choice Experiments

Beavers readily respond to food provided at their pond's edge. This has permitted many experiments with standard sizes and numbers of sticks or twigs. Such experiments vary in their presentation: the food can be laid on the ground, stuck into the soil, fastened to some base, or placed in the water. With luck, it is possible to directly observe the beavers' choices. Plates 24 and 25 show a beaver sniffing, selecting, and then carrying away a bundle of experimental sticks that had been laid on the pond's bank. When the natural foraging behavior of beavers is mimicked by fastening the experimental sticks to a holder, the animals will bite off desirable samples. Figure 9.6 and Plate 26 show arrays of sticks before and after the beavers made their choices.

REFERENCES

1. Svendsen, G. E. 1980. Seasonal change in feeding patterns of beaver in southeastern Ohio. Journal of Wildlife Management 44: 285–290.
2. Brenner, F. J. 1962. Foods consumed by beavers in Crawford County, Pennsylvania. Journal of Wildlife Management 26: 104–107.
3. Jenkins, S. H. 1979. Seasonal and year-to-year differences in food selection by beavers. Oecologia 44: 112–116.

4. Boyce, M. S. 1981. Habitat ecology of an unexploited population of beavers in interior Alaska. In: J. A. Chapman and D. Pursley, editors. Proceedings of Worldwide Furbearer Conference; 1980 Aug. 3–11; Frostburg, Md. Volume 1. Frostburg, Md.: Worldwide Furbearer Conference, Inc. p 155–186.
5. Fryxell, J. M. 1992. Space use by beavers in relation to resource abundance. Oikos 64: 474–478.
6. Müller-Schwarze, D., and B. A. Schulte. 1999. Behavioral and ecological characteristics of a "climax" population of beaver (*Castor canadensis*). In: P. E. Busher and R. M. Dzięciołowski, editors. Beaver protection, management, and utilization in Europe and North America. New York: Kluwer Academic/Plenum. p 161–177.
7. McGinley, M. A., and T. G. Whitham. 1985. Central place foraging by beavers (*Castor canadensis*): a test of foraging predictions and the impact of selective feeding on the growth of cottonwoods (*Populus fremontii*). Oecologia 66: 558–562.
8. Belovsky, G. E. 1984. Summer diet optimization by beaver. American Midland Naturalist 111: 209–222.
9. Hall, J. G. 1960. Willow and aspen in the ecology of beaver on Sagehen Creek, California. Ecology 41: 484–494.
10. Shadle, A. R., and T. S. Austin. 1939. Fifteen months of beaver work at Allegany State Park, N. Y. Journal of Mammalogy 20: 299–303.
11. Henry, D. B., and T. A. Bookhout. 1970. Utilization of woody plants by beavers in northwestern Ohio. Ohio Journal of Science 70: 123–127.
12. O'Brien, D. F. 1938. A qualitative and quantitative food habit study of beavers in Maine [M.S. thesis]. Orono: University of Maine.
13. Doucet, C. M., and J. M. Fryxell. 1993. The effect of nutritional quality on forage preference by beavers. Oikos 67: 201–208.
14. Müller-Schwarze, D., B. A. Schulte, L. Sun, A. Müller-Schwarze, and C. Müller-Schwarze. 1994. Red maple (*Acer rubrum*) inhibits feeding by beaver (*Castor canadensis*). Journal of Chemical Ecology 20: 2021–2034.
15. Shelton, P. C. 1966. Ecological studies of beavers, wolves, and moose in Isle Royal National Park, Michigan [Ph.D. dissertation]. Lafayette, Ind.: Purdue University.
16. Jenkins, S. H. 1978. Food selection by beavers: sampling behavior. Breviora no. 447: 1–6.
17. Doucet, C. M., R. A. Walton, and J. M. Fryxell. 1994. Perceptual cues used by beavers foraging on woody plants. Animal Behaviour 47: 1482–1484.
18. Jenkins, S. H. 1980. A size-distance relation in food selection by beavers. Ecology 61: 740–746.
19. Jenkins, S. H. 1981. Problems, progress and prospects in studies of food selection by beavers. In: J. A. Chapman and D. Pursley, editors. Proceedings of Worldwide Furbearer Conference; 1980 Aug. 3–11; Frostburg, Md. Volume 1. Frostburg, Md.: Worldwide Furbearer Conference, Inc. p 559–579.
20. Fryxell, J. M., and C. M. Doucet. 1993. Diet choice and the functional response of beavers. Ecology 74: 1297–1306.
21. Robbins, C. T. 1983. Wildlife feeding and nutrition. New York: Academic.
22. Basey, J. M., S. H. Jenkins, and P. E. Busher. 1988. Optimal central-place foraging by

beavers: tree-size selection in relation to defensive chemicals of quaking aspen. Oecologia 76: 278–282.
23. Basey, J. M., S. H. Jenkins, and P. E. Busher. 1990. Food selection by beavers in relation to inducible defenses of *Populus tremuloides.* Oikos 59: 57–62.
24. Hagermann, A. E. and C. T. Robbins. 1993. Specificity of tannin-binding salivary proteins relative to diet selection by mammals. Canadian Journal of Zoology 71: 628–633.
25. Bryant, J. P., and P. J. Kuropat. 1980. Selection of winter forage by subarctic browsing vertebrates: the role of plant chemistry. Annual Review of Ecology and Systematics 11: 261–285.
26. Müller-Schwarze, D., H. Brashear, R. Kinnel, K. A. Hintz, A. Lioubomorov, and C. Skibo. 2001. Food processing by animals: do beavers leach tree bark to improve palatability? Journal of Chemical Ecology 27: 1011–1026.
27. Doucet, C. M., and J. P. Ball. 1994. Analysis of digestion data: apparent and true digestibilities of foods eaten by beavers. American Midland Naturalist 132: 239–247.
28. Currier, A., W. D. Kitts, and I. McT. Cowan. 1960. Cellulose digestion in the beaver (*Castor canadensis*). Canadian Journal of Zoology 38: 1109–1116.
29. Buech, R. R. 1984. Ontogeny and diurnal cycle of fecal reingestion in the North American beaver (*Castor canadensis*). Journal of Mammalogy 65: 347–350.
30. Baker, B. W., and B. S. Cade. 1995. Predicting biomass of beaver food from willow stem diameters. Journal of Wildlife Management 48: 322–326.
31. Stegeman, L. C. 1954. The production of aspen and its utilization by beaver on the Huntington Forest. Journal of Wildlife Management 18: 348–358.
32. Aleksiuk, M. 1970. The seasonal food regime of Arctic beavers. Ecology 51: 264–270.
33. Novakowski, N. S. 1967. The winter bioenergetics of a beaver population in northern latitudes. Canadian Journal of Zoology 45: 1107–1118.
34. Slough, B. G. 1978. Beaver food cache structure and utilization. Journal of Wildlife Management 42: 644–646.
35. Busher, P. E. 1991. Food caching behaviour by the North American beaver, *Castor canadensis*, in western Massachusetts. In: Transactions of the 18th International Union of Game Biologists Congress, Krakow 1987. Krakow-Warszawa: Swiat Press. p 111–114.
36. Busher, P. E. 1996. Food caching behaviors of beavers (*Castor canadensis*): selection and use of woody species. American Midland Naturalist 135: 343–348.
37. Brenner, F. J. 1962. Foods consumed by beavers in Crawford County, Pennsylvania. Journal of Wildlife Management 26: 104–107.
38. Roberts, T. S. 1937. How two young beavers constructed a food pile. Proceedings of the Minnesota Academy of Science 5: 24–27.

Part III

Populations

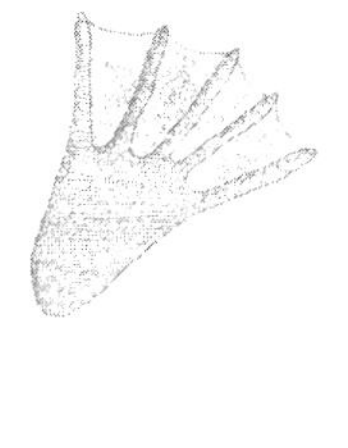

chapter

10 Reproduction, Development, and Life Expectancy

... ascertaining the usual number of their offspring. I have seen some hundreds of them killed at the seasons favourable for those observations and never could discover more than six young in one female, and that only in two instances; for the usual number, as I have before observed, is from two to five.

Samuel Hearne, 1795

Reproduction

Beavers form permanent breeding pairs and are socially, if not genetically, monogamous, as far as we know, even for consecutive breeding seasons. In a colony, only the adult pair breeds. In the Eurasian beaver, females come into estrus from late December on. The peak occurs in mid-January. Warm-weather spells in winter can accelerate estrus. Mating occurs in the water, with the male approaching the floating female from the side. In the Eurasian beaver, mating may also take place in the lodge.[1] Copulation lasts from 30 seconds to 3 minutes. In one study, all successful matings took place from December 30 to May 2, 54% of these in mid and late January. Females were in estrus for 12–24 hours. Ninety-three percent of the copulations took place in the evening or at night. Of 61 pregnant females that were studied, only 10 were fertilized during their first estrus of the season. If not impregnated, females came into estrus 2–4 times per season, with irregular intervals ranging from 7 to 57 days. The estrus cycles, as determined by vaginal smears, lasted from 7 to 12 days.[1]

Gestation time in two captive *C. canadensis* females was 104 and 111 days[2]; it should be 128 days according to Asdell.[3] The nipples of North American beavers enlarge in the second month of gestation. Three or four kits are born between May and July once every year.[4] In *C. fiber* gestation time is about the same, 107 days.[1,5] Also in the European beaver, most young were delivered in early May, but some also from mid-April to mid-August. In the same species, 90% of the births occurred during the daytime, between 1000 and 1400 hours, and they lasted from a few minutes to 2 hours.[5]

In a study in Newfoundland the mean litter size of North American beavers was 2.7 embryos.[6] In Maine there were 3.55 young in every litter, on average.[7] In the same population, each female had an average of 3.95 placental scars, each in-

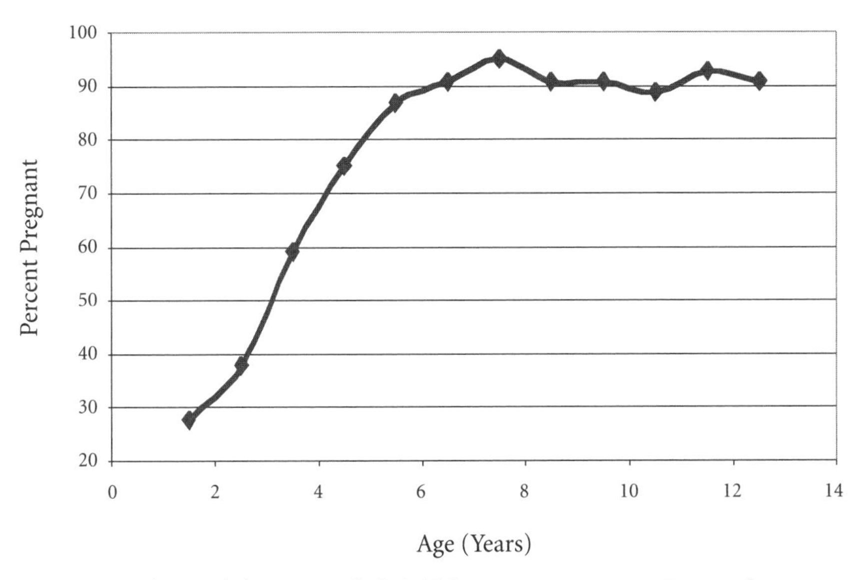

Figure 10.1 | Female beavers reach their highest pregnancy rate at 7 years of age. (After data from Payne.[11])

dicating one embryo in the preceding gestation period. Thus, infant mortality seems to be low, and placental scars are representative of actual litter size. When a large amount of preferred food is available and the weather is mild, one mother can give birth to as many as 6 kits.[7]

The average litter size in captive Eurasian beavers in Poland was 2.7.[5] Females older than 9 years often do not bear young in one particular year but still have large litter sizes, although their young do not survive well.[5] One female on a beaver farm in Poland bore 49 young in 13 litters, for an average litter size of 3.8.[5] Large mothers are more likely to give birth to heavier kits than are small females.[8,9]

Fertility is higher in North American beavers than their Eurasian counterpart. In the Eurasian beaver, only 7%–8% of females reproduce at the age of 1.5–2.0 years, and 50%–60%, as adults. However, 20% of the 1.5- to 2.0-year-old female and 70%–80% of the adult North American beavers reproduce.[10] Among females of captive Eurasian beavers in Poland, 29% gave birth for the first time when 2 years old; 26%, at 3 years; 35%, at 4 years; and 10%, when 5 years old.[4]

From 1.5 to 6.5 years of age, the number of pregnant females in each age class grows steadily. Among 7-year-old females, 90% are pregnant. This rate remains the same until 12.5 years (Fig. 10.1). At that age, few females survive. Payne[11] found only one 12-year-old female in his study, while there were seventy 1.5-year-olds.[11]

Development and Growth

The young are born fully furred and teethed, and the eyes open at birth or several hours later, sometimes after being rubbed with the front paws.[12,13] Each weighs 340–630 g (0.75–1.39 lb). The newborn explore their surroundings and may enter the water the day they are born. Within hours, they hiss and gnash their teeth, defensive behaviors.[14] These characteristics make the beaver an extremely *precocial* animal, defined as an animal that is never helpless and confined to a parental nest, like ducklings, for instance. However, beavers also show characteristics of *altricial* animals: the newborn lack bony epiphyses in the arms and legs. Altricial animals, such as songbirds and humans, are helpless and totally dependent on parental care. Finally, being confined inside the lodge for the first 4–5 weeks, the young beavers lead a *nidicolous* way of life, comparable to that of gulls and penguins, a status somewhat intermediate between being precocial and altricial.

Initially, the fur is not water-repellent. The anal glands, used in greasing the fur, function at the age of 3–4 weeks. After 1.5–2.0 months, the fur is completely water-repellent. Before that time, grooming by the parents most likely provides fat for the fur of the kits. According to observations in one study, hand-reared kits attempted to dive for the first time when 6–7 days old, dove to escape at 8–10 days, were able to dive and swim underwater at 9–10 days, and were fully capable of diving and staying submerged when 2 months old. Scent-marking movements appeared at 13–14 days and the tail slap, at 3–4 weeks. They walked bipedally, galloped, and hopped at 2 months, and gathered construction material while walking by 90 days.[15]

At 1 month, hand-reared kits drank 25 ml of milk per day, their peak consumption. A feeding session lasts from 3 to 15 minutes. During that time, lactating females have very visible, enlarged nipples (Plate 27).

The young sample solid food at 6–7 days and try to gnaw wood at about 11 days. They manipulate food with their paws within 12–14 days and eat plants at 16–18 days, predominantly so when 3 weeks old. Kits start to practice coecotrophy (reingestion of feces) within 3 weeks.[15] Lactation lasts 6–8 weeks,[16,17] but Patenaude and Bovet[18] observed weaning at 10 weeks. From then on, the young live completely on tender grass, forbs, and leaves brought by the parents. Eurasian beavers in captivity lactated for 90 days.[4] The young are nursed during the daytime and sleep during the night when the mother is active outside the lodge.

One-day-old young already groom their face, on day 7 they groom their chest and abdomen, and during week 3 they are able to sit on their tail while grooming themselves[15] (Plate 28). Mutual grooming between kits and between the male and the kits starts at 2 weeks, and wrestling when they are 3 weeks old. Yearlings and kits groom each other from 3 weeks on, and finally, mother and young do so when the latter are 4 weeks old.

Parents and juveniles will rescue a kit that falls in the water inside the lodge. They take the kit into their teeth, push it with their nose, or carry it between forepaw and chin while walking bipedally.[14]

When kits weigh about 3–4 kg (6.6–8.8 lb), they start to leave the lodge regularly to explore the outside world and feed side by side with their parents and older siblings. The growth rate varies among different geographical areas. In New York, kits can reach 6–8 kg (13.2–17.6 lb) by October. They already have stored much fat by then, which is very important for their ability to survive the impending harsh northern winter with its often insufficient food supply. Despite enjoying the shelter of the lodge, kits often do not survive their first winter, most likely because they failed to store enough fat. Predation also takes its toll. In the following spring (May), the now 1-year-olds typically weigh 8.6–10.4 kg (19–23 lb). How fast they grow depends on the food resources. We occasionally found yearlings weighing 6–7 kg (13–15 lb) on poor sites and 11–12 kg (25–26 lb) on better sites at the same time. During their second spring, these beavers reach 11–13 kg (24–29 lb) and are ready for dispersal.

Growth rate slows down with age. Three-year-old beavers typically weigh 13.6–17.0 kg (30–37 lb), and 4-year-olds reach 13.6–20.0 kg (30–44 lb). Afterward, the body weight fluctuates with the seasons. Beavers lose weight over the winter and gain again during the growing season (Fig. 10.2). The largest beaver we have trapped was a 26.2-kg (58-lb) adult female. The beaver was so big that it almost filled the entire space of our Hancock live trap.

Removal by fur trappers may change some life-history aspects of beavers. In Alaska, female beavers in a population that is regularly trapped reproduce at a younger age, are smaller when mature, and suffer a higher mortality compared to those in a population that remains intact.[19]

Life Expectancy

Although a 21-year-old North American beaver has been reported,[20] most beavers live only about 10 years.[21] Their life span in the wild is abut 12 years, and in captivity they have lived for 19 years. In our own studies (unpublished) the oldest beaver was 12 years old at its death.

In a sample of Eurasian beavers, the oldest one was 18 years old.[22] Several anatomical measures indicate the age of a beaver. These include size and proportions of the skull, the closing of the epiphysis of the femur bone (very visible up to 30 months, fully fused with 7 years), annual layers of cementum in the molar and premolar teeth, and the weight and proportions of the *baculum*, or "penis bone."[23] The gradual closing of the open base of the molars from 1.5 to 4.5 years of age is also useful for aging beavers.[24] All of these have been measured on dead animals.

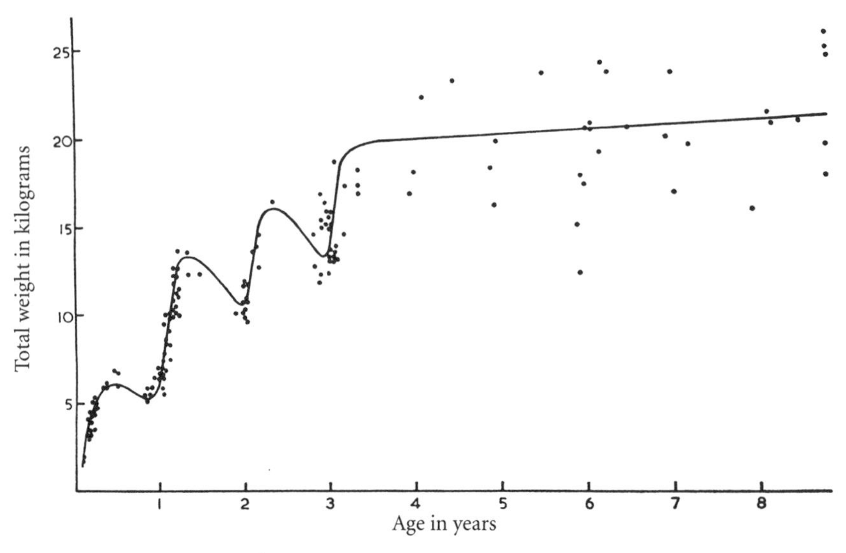

Total weight growth curve to 8 years of age.

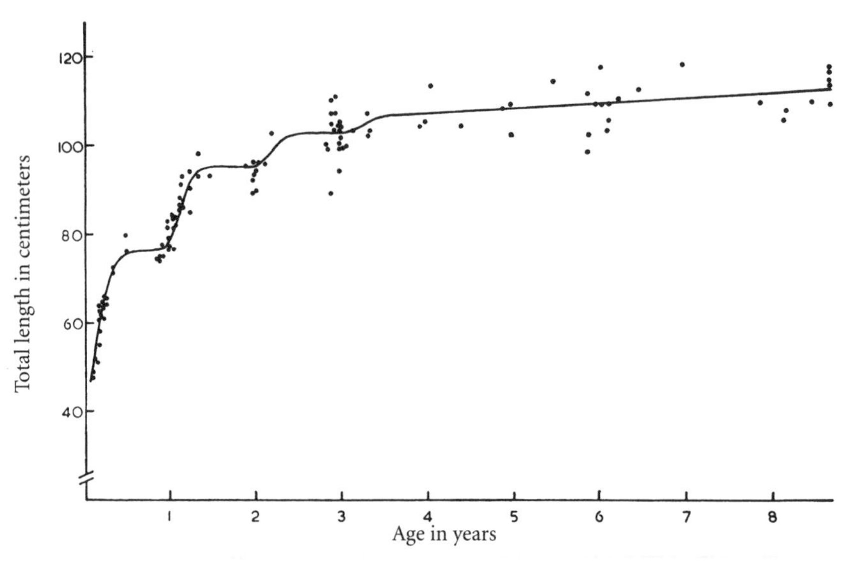

Skeletal (total length) growth curve to 8 years of age.

Figure 10.2 | Growth curves of Makenzie Delta beavers to 8 years of age. Top: Body weights. Bottom: Body lengths. From M. Aleksiuk and I. McT. Cowan. 1969. Canadian Journal of Zoology 14: 474, Fig. 3. Reproduced with permission from NRC Research Press.

Pair Bonds

When about 23 months old, the young leave the parental colony. While still with the parents, they help to raise siblings but will not breed themselves. If one member of the breeding pair is lost, the other will breed with offspring or immigrants.[1,25] At Allegany State Park we observed several cases of mate replacement: an adult from another family replaced an adult of the same sex. However, we still do not know exactly how mate replacement happens.

Sterilization of the adult breeding male or female does not trigger reproduction by the rest of the colony, as long as both mates remain in the colony. Only after the sterilized individual moves out, a 2-year-old family member forms a new pair, and reproduction in the colony resumes.[26]

Mate replacement occurs relatively often. In Ohio, the average pair bond lasts 2.5 years, although some pairs may remain together for 8 years.[25] In California, the pair bond lasts on average 3.1 years.[27] Most new pairs involve one 2-year-old disperser. This age difference between the mates may contribute to the high replacement rate because after the older partner dies, the younger one has to form another pair bond. An adult beaver that has lost its mate will usually stay on the same site, waiting for a new suitable mate to arrive. Very rarely are two females in the same lodge pregnant at the same time. One such case occurred in southern Manitoba, Canada.[28]

Habitat and Reproduction

Differences in the quality of habitat can affect the reproductive effort of beavers. Put another way, these animals are able to adjust their reproduction to environmental conditions. For instance, in northern Saskatchewan, beavers in predominantly coniferous forest, considered poor habitat, produced fewer young than their counterparts in mixed forests.[29] Table 10.1 lists some differences in reproduction in poor and good habitats.

Table 10.1 | Differences of Reproduction by Beavers in Good and Poor Habitats in Northern Saskatchewan

Variable	In Good Habitat (Mixed Forest)	In Poor Habitat (Coniferous Forest)
Ovulation rate (number of corpora lutea)	5.20	3.00
Reabsorption rate (%)	1.8	13.8
Pregnancy rate (%)	85	71.4
Productivity (% kits)	51.5	23.4

Source: Reference 29.

REFERENCES

1. Wilsson, L. 1971. Observation and experiments on the ethology of the European beaver (*Castor fiber* L.). Viltrevy 8: 115–266.
2. Hediger, H. 1970. The breeding behavior of the Canadian beaver (*Castor fiber canadensis*). forma et functio 2: 336–351.
3. Asdell, S. A. 1964. Patterns of mammalian reproduction. 2nd ed. Ithaca: Cornell University Press (Comstock).
4. Hodgdon, K. W., and J. H. Hunt. 1953. Beaver management in Maine. Maine Department of Inland Fisheries and Game, Game Division Bulletin 3: 1–102.
5. Doboszyńska, T., and W. Zurowski. 1983. Reproduction of the European beaver. Acta Zoologica Fennica 174: 123–126.
6. Bergerud, A. T., and D. R. Miller. 1977. Population dynamics of Newfoundland beaver. Canadian Journal of Zoology 55: 1480–1492.
7. Hodgdon, K. W. 1949. Productivity data from placental scars in beavers. Journal of Wildlife Management 13: 412–414.
8. Pearson, A. M. 1960. A study of the growth and reproduction of the beaver (*Castor canadensis* Kuhl) correlated with the quality and quantity of some habitat factors [M.S. thesis]. Vancouver: University of British Columbia.
9. Boyce, M. S. 1974. Beaver population ecology in interior Alaska [M.S. thesis]. College: University of Alaska.
10. Danilov, P. I., and V. Ya. Kan'shiev. 1983. The state of population and ecological characteristics of European (*Castor fiber* L.) and Canadian (*Castor canadensis* Kuhl.) beavers in the northwestern USSR. Acta Zoologica Fennica 174: 95–97.
11. Payne, N. F. 1984. Reproduction rates of beaver in Newfoundland. Journal of Wildlife Management 48: 912–917.
12. Bradt, G. W. 1939. Breeding habits of beaver. Journal of Mammalogy 20: 486–489.
13. Guenther, S. E. 1948. Young beavers. Journal of Mammalogy 29: 419–420.
14. Patenaude, F. 1983. Care of young in a family of wild beavers, *Castor canadensis*. Acta Zoologica Fennica 174: 121–122.
15. Lancia, R. A., and H. E. Hodgdon. 1983. Observations on the ontogeny of behavior of hand-reared beavers (*Castor canadensis*). Acta Zoologica Fennica 174: 117–119.
16. Bailey, V. 1927. Beaver habits and experiments in beaver culture. Technical Bulletin of U.S. Department of Agriculture 21: 1–40.
17. Jenkins, S. H., and P. E. Busher. 1979. *Castor canadensis*. Mammalian Species 120: 1–8.
18. Patenaude, F., and J. Bovet. 1983. Parturition and related behavior in wild American beavers (*Castor canadensis*). Zeitschrift für Säugetierkunde 48: 136–145.
19. Boyce, M. S. 1981. Beaver life-history responses to exploitation. Journal of Applied Ecology 18: 749–753.
20. Larson, J. S. 1967. Age structure and sexual maturity within a western Maryland beaver (*Castor canadensis*) population. Journal of Mammalogy 48: 408–413.
21. Novak, M. 1977. Determining the average size and composition of beaver families. Journal of Wildlife Management 41: 751–754.

22. Stiefel, A. and R. Piechocki. 1986. Circannuelle Zuwachslinien im Molarenzement des Bibers (*Castor fiber*) als Hilfsmittel für exakte Altersbestimmungen. Zoologische Abhandlungen, Staatliches Museum für Tierkunde, Dresden 41: 165–175.
23. Piechocki, R. 1986. Osteologische Kriterien zur Altersbestimmung des Elbebibers *Castor fiber albicus*. Zoologische Abhandlungen, Staatliches Museum für Tierkunde, Dresden 41: 177–183.
24. Van Nostrand, F. C., and A. B. Stephenson. 1964. Age determination for beavers by tooth development. Journal of Wildlife Management 28: 430–434.
25. Svendsen, G. E. 1988. Pair formation, duration of pair-bonds, and mate replacement in a population of beavers (*Castor canadensis*). Canadian Journal of Zoology 67: 336–340.
26. Brooks, R. P., M. W. Fleming, and J. J. Kenelly. 1980. Beaver colony response to fertility control: evaluating a concept. Journal of Wildlife Management 44: 568–575.
27. Taylor, D. 1970. Growth, decline, and equilibrium in a beaver population at Sagehen Creek, California [Ph.D. dissertation]. Berkeley: University of California.
28. Wheatley, M. 1993. Report of two pregnant beavers, *Castor canadensis*, at one beaver lodge. Canadian Field-Naturalist 107: 103.
29. Gunson, J. R. 1970. Dynamics of the beaver of Saskatchewan's northern forest [M.S. thesis]. Edmonton: University of Alberta.

chapter

11 Population Densities and Dynamics

Human exploitation of beavers decreases the survivorship of adults, but by freeing high quality colony sites, results in enhanced survivorship for dispersing pre-reproductives. Females breeding earlier in life in an exploited population attain smaller size at maturity and consequently suffer higher mortality than individuals breeding later at a larger body size. These trade-offs between fecundity, growth and survivorship are as predicted by recent theory on the evolution of life-histories.

M. S. Boyce, 1981

Population Densities

How many beavers can live in a given area? Because beavers hold territories that contain essential food and water resources, their population density in a given area is limited. Water is indispensable to beavers; therefore, the density of beavers is traditionally calculated as the number of colonies along a unit length of stream and the numbers of beavers in each colony. Suitable habitats can accommodate up to 1.2 colonies/km of stream (1.8/mile).[1–5] Nevertheless, others calculate the population density as number of colonies per area.

In Canada, beaver densities of untrapped populations in the western mixed forests of the national parks were 1.06 and 1.18 colonies/km.[2,7] By contrast, in New Brunswick trapped populations averaged only 0.33 colony/km^2, but untrapped ones 1.06 colonies. An unusually high density of 3.51 colonies/km^2 occurred in central Alberta, which resulted in problems caused by "nuisance beavers."[7] The maximum average density for Canada as a whole was estimated at 1.0–1.2 colonies/km^2.[7] Table 11.1 lists some beaver densities as colonies per mile and km stream in a number of North American States or Provinces.

Colony Size

The average number of beavers living in a colony ranges from 4 in western New York and Alaska[2] to over 8 in Massachusetts[8] and Nevada.[9] Colony sizes found in various studies are summarized by Müller-Schwarze and Schulte.[6] See also table 11.2.

Table 11.1 | Beaver Densities (Number of Colonies per Unit Stream Length) in Various Areas

Area	No./mile	No./km
Alaska	0.64	0.40
Fulton County, New York	0.87	0.54
Massachusetts	0.89	0.55
Western New York	0.93	0.58
Eastern South Dakota	1.30	0.81
Quabbin Reservation, Mass.	1.61	1.00
New Brunswick, N.J.	1.76	1.09

Source: Modified from reference 6.

Table 11.2. | Numbers of Beavers per Family in Various Areas

Area	Average No. / Family
Alaska	4.1
Montana	4.1
Newfoundland	4.2
Adirondacks, N.Y.	4.3
Sagehen Creek, Calif.	4.8
Michigan	5.1
Allegany State Park, N.Y.	5.4
Ohio	5.9
Colorado	6.3
Isle Royale National Park, Mich.	6.4
Massachusetts	8.1
Nevada	8.2

Source: Reference 6.

Population Composition

In a study in South Dakota, the age classes were represented as follows: There were 19.5% kits, 22.6% yearlings, and 57.9% adults, defined as 2.5 years or older. For all age classes, males were slightly more numerous (male-female ratio: 1.24 : 1.00), but this difference was not statistically significant. Of the beavers aged 4.5 years and older, 66% were males, a significant sex difference.[10] Many possible reasons for higher female mortality could be invoked. The age class composition in different studies proved remarkably similar (Fig. 11.1).

On the other hand, kits and yearlings can constitute quite different proportions of the same population, depending on the phase of the population cycle. For example, at the Quabbin Reservation in Massachusetts, these age classes rep-

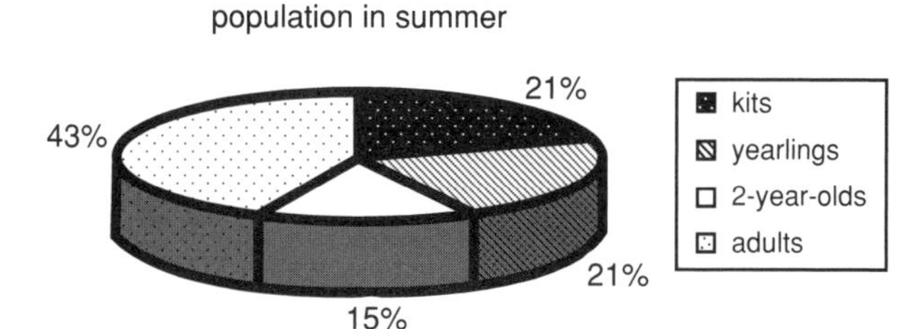

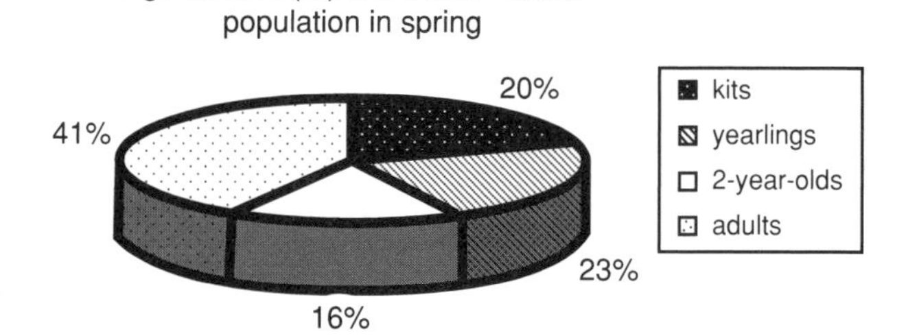

Figure 11.1 | Percentages of age classes in different populations can be very similar. (After data from references 3 and 10.)

resented a large and fairly constant percentage of all beavers during the growth phase, whereas they were less frequent when the population stabilized after a steep decline[11] (Fig. 11.2).

Survivorship Curve

In populations where the exact age in years is known for a large number of individuals, a survivorship curve can be developed. As the beavers grow older, fewer survive. The first years of life see rapid attrition: yearlings represent at least 30% of the population, but 4-year-olds constitute less than 10%[12] (Fig. 11.3).

Population Dynamics

Population Growth

Beaver populations change slowly and lack the boom-and-bust pattern that small rodents undergo. The size of an undisturbed population is regulated mainly by how much suitable habitat is available. Each family produces about three kits and contributes about two per year to the population. If three kits are born, none die, and they start breeding within 3 years, the population would grow to 278 in 10 years (Fig.11.4). But young beavers suffer high mortality due to severe weather, lack of preferred food, and predation from carnivores. Wolves, coyotes, wolverines, bears, and even mink can depredate beavers, especially yearlings.[13–17] Year-

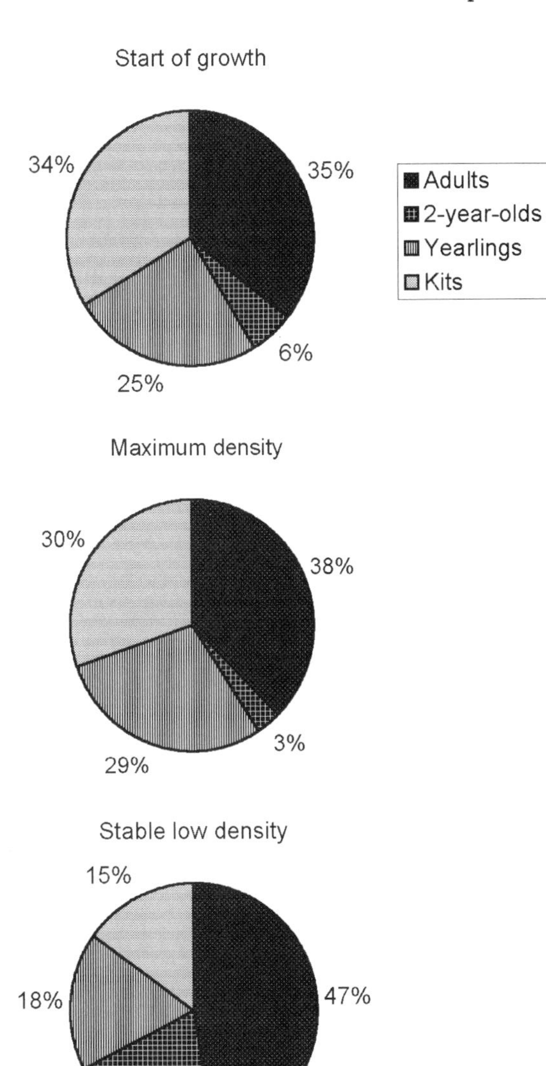

Figure 11.2 | Percentages of age classes during different phases of population cycle: growing population, maximum density, and stable low density after decline. Note the relative increase in adults and 2-year-olds and lower number of kits and yearlings in stable low density population. (After data from Busher and Lyons.[11])

lings constitute about one-third of a population. As many as 30% of the beavers of all age classes can die in 1 year in Newfoundland.[3]

The beaver population in Allegany State Park illustrates how it changes over time. Trappers had eliminated all beavers around the start of the 20th century. In 1937 a pair of beavers from the Adirondacks was introduced, and 1 year later two families were living in the park. Beavers occupied virtually the entire suitable

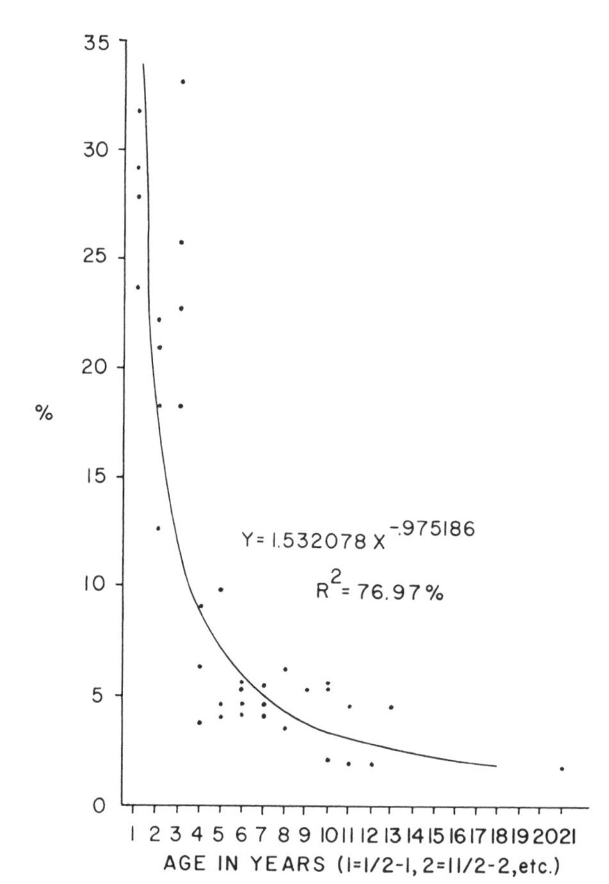

Figure 11.3 | Survivorship curve for Maryland beavers. (Reproduced from J. S. Larson. 1967. Age structure and sexual maturity within a western Maryland beaver (*Castor canadensis*) population. Journal of Mammalogy 48: 408–413.)

habitat by the 1950s. In 1973, the now 34 beaver families residing in the park caused considerable damage. The park managers responded by removing 93 beavers. Only 14 families remained. Since then, the population grew by 3–4 new families every year and reached 60 in the mid-1980s. From then on, the population slightly dropped to 40–60 families, probably because suitable habitat and food became scarce. The dynamics of another beaver population in Newfoundland followed almost the same pattern[3] as our Allegany population. Busher[5] followed the growth of a third well-studied population in California. Within 25 years, the population expanded from occupying 20% of suitable stream habitat to occupying 56%. The number of beavers per kilometer of stream grew from 1.57 to 4.00. But between these two stages the population also declined temporarily (Table 11.3). The number of beaver colonies at the Quabbin Reservation in Massachusetts grew slowly from 2 to 16 over 17 years. After that, the population expanded rapidly, reaching a peak of 54 colonies in 13 more years. A 7-year-long

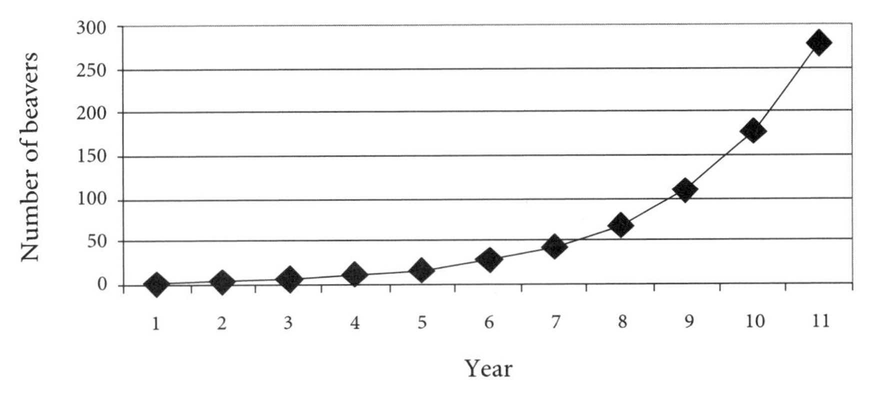

Figure 11.4 | Reproductive potential of beavers: the number of offspring of one pair after 10 years if all beavers survived.

rapid decline followed, until the population stabilized at the low level of 10–15 colonies[11] (Fig. 11.5).

The annual increase was estimated as 49% for beavers of an Ohio population. This rate would permit a 32% annual harvest, and the population would still remain stable.[18]

In Europe, the growth of reintroduced populations is well documented. For example, in Sweden, after extirpation, the first pair was released in 1922. By 1935 the country-wide number was estimated at over 400, and in 1999 it had mushroomed to over 100,000. Norway had approximately 100 beavers in 1899 and about 50,000 in 1999.[19] The Swedish beavers initially experienced population increases of 20%,

Table 11.3 | The Growth of a Beaver Population in California

Year	Stage	Beavers/km	Meter Stream/ Colony	Percentage of Habitable Stream Occupied
1955	Initial colonization	1.57	585	20
1961	Stabilization, high level	3.83	766	67
1968	Decline	0.70	600	10
1974	Secondary expansion	1.57	600	21
1979	Secondary expansion, high level	4.00	800	56

Source: Reference 5.

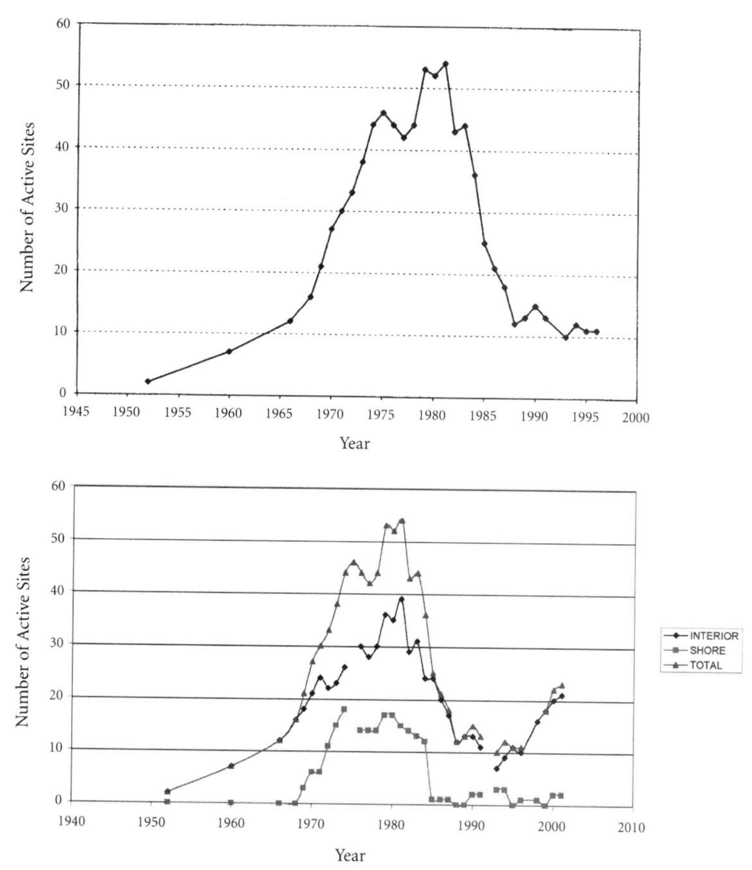

Figure 11.5 | Changes of beaver population sizes over several decades. Top: Sagehen Creek, California. (Reproduced with permission from P. E. Busher and P. J. Lyons. 1999. Long-term population dynamics of the North American beaver, *Castor canadensis*, on Quabbin Reservation, Massachusetts, and Sagehen Creek, California. In: P. E. Busher and R. M. Dzięciołowski, editors. Beaver protection, management, and utilization in Europe and North America. New York: Kluwer Academic/Plenum. p 155, Figure 4.) Bottom: Prescott Peninsula, Quabbin Reservation, Massachusetts. (Courtesy Dr. Peter E. Busher, 2002.)

but the growth rate slowed down to zero about 50 years after reintroduction. The population's rate of increase (r) also slowed down with increasing population density[20] (Fig. 11.6).

In the United States, beavers were absent from the Isle Royale National Park in Michigan during the 19th century. Then fires prepared the ground for an outburst of quaking aspen and white birch. Beavers found this bonanza and were well established by 1920. In 1930, 27 sites were recorded. After reaching a peak in

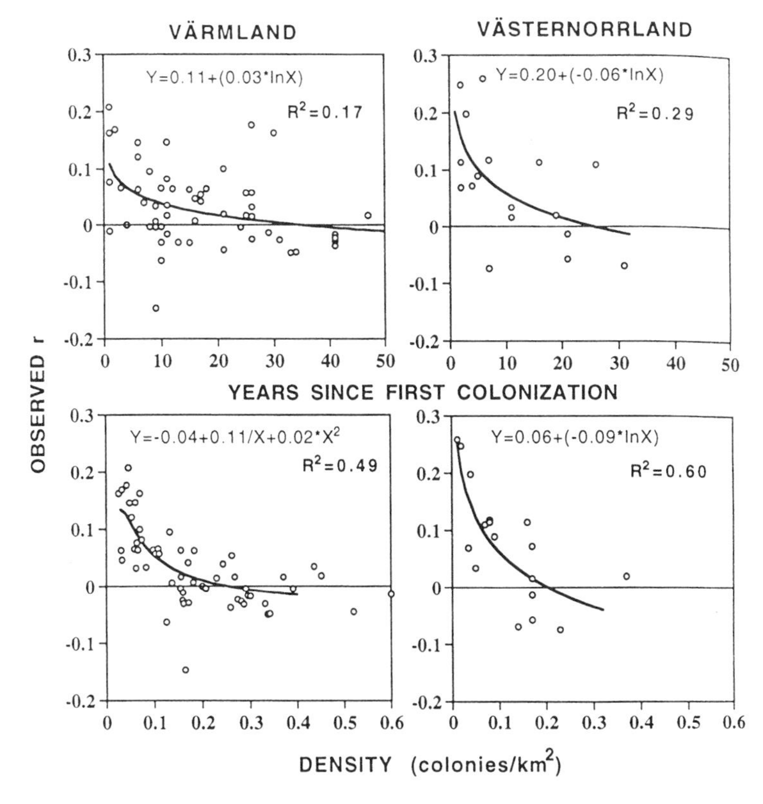

Figure 11.6 | Beaver populations grow fast initially and decelerate their growth rate as the population density increases. (From G. Hartman. 1994. Long-term population development of a reintroduced beaver (*Castor fiber*) population in Sweden. Conservation Biology 8: 716, Figure 3. Reproduced with permission from Blackwell Science.)

1948, the population declined, presumably decimated by an outbreak of tularemia, which breached this area of Michigan in 1953.[21]

Response to Removal of Beavers

Very few beaver populations are "natural." Most are being exploited by fur trappers, or have been in the recent past, or are being actively managed. In many places, beavers have been introduced, as in Tierra del Fuego, or reintroduced, as in various European countries and the northeastern United States. Residential, agricultural, forestry, and transportation uses of the landscape limit expansion of all of these beaver populations.

Experimental removal of beavers from a small area permits researchers to study the recolonization. In Newfoundland, Bergerud removed "nearly all the beaver" from 14 colonies. One year later, 7 new colonies appeared. The number grew to 12 the next year, and 17 after that, leveling off at 15 during the fourth year, indicating the population had recovered.[3]

Exploitation frees good sites for colonization by young beavers and thus reduces the mortality of 2-year-olds when they disperse from their natal colonies. Heavily exploited populations respond by changing their reproductive patterns.

For instance, many yearlings breed, which speeds up reproduction and population growth. But the females that breed earlier attain a smaller body size when mature. This in turn leads to a higher mortality than what females suffer when they breed later and at a larger size. The unexploited population has fewer 3-year-olds, possibly because 2- to 3-year-olds cannot find homes. Many of this age class die from predation, intraspecific strife, or subsisting in marginal habitat. Thus, beavers face trade-offs between fecundity, body growth, and individual survival.[22] Because of these trade-offs, an exploited and an unexploited population can have the same population density, as Boyce[22] found in Alaska.

Response to Population Saturation

Unexploited populations grow to a saturation point, at the carrying capacity of the habitat, which changes with fluctuating water and food resources. Beavers reaching sexual maturity experience difficulties in finding a suitable site to breed. Under these conditions, many 2-year-olds stay at their parental colony for 1 more year, instead of emigrating. This effect can kick in early during population growth: in New York no yearlings bred when a mere 40% of all potential beaver sites were occupied.[23]

Characteristics of a Climax Population

What will happen to a beaver population if it is allowed to grow unconstrained? The population at Allegany State Park had been neither managed nor trapped for 16 years at the time of our study.[6] By all accounts, it had reached carrying capacity and constituted a "climax population," characterized by no further increase in the number of colonies. This "unexploited" population differed in four ways from "managed" counterparts (the vast majority of populations). In another four ways, the beavers showed no differences.

Differences

1. Stream gradients. Beavers in the dense population colonized steeper streams. The six steepest gradients we could find averaged 4.2%; of these, the two steepest gradients measured 6.4% and 6.3%. By contrast, beavers in Fulton County, New York, where only 30% of the sites were occupied, inhabited gradients that ranged from 0.3% to 0.8% and averaged 0.6%.

2. Stay-home youngsters. A large percentage of 2-year-olds stayed in their parental colony for 1 more year. Of 45 colonies, 17 (37.8%) housed one additional "adult" beyond the breeding pair. All tagged "supernumeraries" proved to be older offspring of the local breeding pair.

3. Feeding on less preferred tree species. Compared to the food the original colonizers chose in 1937–41, the Allegany beavers favored different tree species in

the 1980s, since preferred species had been depleted, at least near their ponds. Specifically, the beavers consumed the preferred *Populus* and *Salix* trees significantly less in 1984/85 than in 1937–41, while they had statistically increased their take of *Prunus* and to a lesser degree *Fagus*.

4. Longer trails. Upon depletion of preferred trees around their ponds, beavers had to go farther for aspen, for example. "Regular" trails, found at all colonies with trees growing near the water, ranged from about 15 to 80 m in length. One sample of 18 trails averaged 31.5 m. If, however, the area around the pond was completely depleted of trees, beavers foraged at greater distances (moving to another location is not an option because of the high population density). Six of such trails to distant aspen groves averaged 129 m and ranged from 65 to 201 m.

Similarities

1. Population density. The overall population density—0.52 colony/stream kilometer—was not significantly different from densities in other areas from various studies. However, the density of colonies varied from stream to stream. One stream reached one of the highest densities ever reported, at 1.14 colonies/km, while we found only 0.2 colony/km at another.
2. Body weight. The beavers at Allegany State Park did not weigh less or more than those in other areas, both in New York and elsewhere.
3. Family size. The number of family members ranged from 4.1 to 6.12 in different years. Other studies found 4.1–8.2 beavers/colony.[5,9,21] Therefore, family sizes in this "climax" population lie in the lower and middle ranges of other populations. They are similar to family sizes of northern populations, such as in Alaska, Montana, and Newfoundland, but smaller than those in Massachusetts, Michigan, and Nevada.
4. Numbers of scent mounds and response to scent. We don't have much comparative data on the numbers of scent mounds and the beavers' responses to experimental scent marks. However, the number of scent mounds at a given colony is correlated with the number of colonies nearby (i.e., the local population density).

Like other mammals such as elephant seals, beavers respond to a population decline by accelerating reproduction. This is accomplished by younger females bearing young. Under such conditions, a higher percentage of 2-year-old females, even yearlings, become pregnant.

Continent-wide Numbers

On a continent-wide scale, an estimated 60–400 million beavers may have lived in North America before Europeans arrived. Such estimates mean 6–37 beavers/stream kilometer, or 10–60 animals/mile.[24] In 1909, Seton-Thomson[25] estimated the total original beaver population to have been perhaps 18 million, certainly "not less than 10,000,000 in years of abundance." Taking into account the population density of 50 beavers to the square mile at Algonquin Park in Canada, reported to him by Dr. T. S. Palmer, Seton-Thomson suggested in a footnote to multiply his earlier estimate by 5. This results in 50–90 million beavers over their entire range. He based his estimates on the annual take of beaver pelts by Hudson's Bay Company, assuming that a beaver population is sustainable if 20% is taken each year. So he multiplied the pelt harvest by 5 to arrive at a population estimate. At the height of the fur trade between 1860 and 1870, the American fur companies and Hudson's Bay Company brought out 153,000 beaver pelts per annum. Adding what the natives used for themselves, the number caught would have been about 500,000. Since other agencies also trapped beavers, the total take was probably around 1 million. Assuming a 20% catch, the total number of beavers in Canada might have been 5 million at that time. For the Adirondack Mountains alone, the French explorer Samuel de Champlain estimated 1 million beavers at the beginning of the 17th century.[26] Although he overestimated the numbers, the woods must have been thick with beavers, a density not seen anywhere today.

REFERENCES

1. Nordstrom, W. R. 1972. Comparison of trapped and untrapped beaver populations in New Brunswick [M.S. thesis]. Fredericton: University of New Brunswick.
2. Boyce, M. S. 1981. Habitat ecology of an unexploited population of beavers in interior Alaska. In: J. A. Chapman and D. Pursley, editors. Proceedings of Worldwide Furbearer Conference; 1980 Aug. 3–11; Frostburg, Md. Volume 1. Frostburg, Md.: Worldwide Furbearer Conference, Inc. p 155–186.
3. Bergerud, A. T., and D. R. Miller. 1976. Population dynamics of Newfoundland beaver. Canadian Journal of Zoology 55: 1480–1492.
4. Allen, A. W. 1983. Habitat suitability index models: beaver. Washington, D.C.: U.S. Department of the Interior, Fish and Wildlife Service. p 1–20.
5. Busher, P. E. 1987. Population parameters and family composition of beaver in California. Journal of Mammalogy 68: 860–864.
6. Müller-Schwarze, D., and B. A. Schulte. 1999. Behavioral and ecological characteristics of a "climax" population of beaver (*Castor canadensis*). In: P. E. Busher and R. M. Dzięciołowski, editors. Beaver protection, management, and utilization in Europe and North America. New York: Kluwer Academic/Plenum. p 161–177.
7. Larson, J. S., and J. R. Gunson. 1983. Status of the beaver in North America. Acta Zoologica Fennica 174: 91–93.

8. Brooks, R. P., M. W. Fleming, and J. J. Kennelly. 1980. Beaver colony responses to fertility control: evaluating a concept. Journal of Wildlife Management 44: 568–575.
9. Busher, P. E., R. J. Warner, and S. H. Jenkins. 1983. Population density, colony composition, and local movements in two Sierra Nevada beaver populations. Journal of Mammalogy 64: 314–318.
10. Dieter, C. D. 1992. Population characteristics of beavers in eastern South Dakota. American Midland Naturalist 128: 191–196.
11. Busher, P. E., and P. J. Lyons. 1999. Long-term population dynamics of the North American beaver, *Castor canadensis*, on Quabbin Reservation, Massachusetts, and Sagehen Creek, California. In: P. E. Busher and R. M. Dzięciołowski, editors. Beaver protection, management, and utilization in Europe and North America. New York: Kluwer Academic/Plenum. p 147–160.
12. Larson, J. S. 1967. Age structure and sexual maturity within a western Maryland beaver (*Castor canadensis*) population. Journal of Mammalogy 48: 408–413.
13. Voigt, D. R., G. B. Kolenosky, and D. H. Pimlott. 1976. Changes in summer food of wolves in Central Ontario. Journal of Wildlife Management 40: 663–668.
14. Young, S. P., and H. H. T. Jackson. 1951. The clever coyote. Harrisburg, Pa.: Stackpole.
15. Rausch, R. A., and A. M. Peterson. 1972. Notes on the wolverine in Alaska and the Yukon Territory. Journal of Wildlife Management 36: 249–268.
16. Hakala, J. B. 1952. The life history and general ecology of the beaver (*Castor canadensis* Kuhl) in interior Alaska [M.S. thesis]. College: University of Alaska.
17. Swank, W. G. 1949. Beaver ecology and management in West Virginia. Conservation Commissioners of West Virginia, Division of Game Management, Bulletin 1. p 1–65.
18. Henry, D. B., and T. A. Bookhout. 1969. Productivity of beavers in northeastern Ohio. Journal of Wildlife Management 33: 927–932.
19. Rosell, F., and K. V. Pedersen. 1999. *Bever.* Aurskog, Norway: Landbruksforlaget.
20. Hartman, G. 1994. Long-term population development of a reintroduced beaver (*Castor fiber*) population in Sweden. Conservation Biology 8: 713–717.
21. Shelton, P. C. 1976. Population studies of beavers in Isle Royale National Park, Michigan. In: R. M. Linn, editor. Proceedings of First Conference of Scientific Research in the National Parks; 1976 Nov. 9–12; New Orleans, La. Volume 1. Washington, D.C.: National Park Service. p 353–356.
22. Boyce, M. S. 1981. Beaver life history responses to exploitation. Journal of Applied Ecology 18: 749–753.
23. Parsons, G. R., and M. Brown. 1979. Yearling production in beaver as related to population density in portions of New York. Transactions of Northeast Fish and Wildlife Conference 16: 180–191.
24. Naiman, R. J., J. M. Melillo, and J. E. Hobbie. 1986. Ecosystem alteration of boreal forest streams by beaver (*Castor canadensis*). Ecology 67: 1254–1269.
25. Seton-Thomson, E. 1909. Life-histories of northern animals; an account of the mammals of Manitoba. 2 volumes. New York: Charles Scribner's Sons.
26. Müller-Schwarze, D. 1992. Beaver waterworks. Natural History 5: 52–53.

chapter 12 Finding a Home: Dispersal

> Often a great colony of many members is lodged in one house. But, if they be incommoded by the narrowness of the place, the younger ones depart of their own accord and construct houses for themselves.
>
> *Father Joseph Jouvency, 1610–13*

Young beavers have to leave their family at one point. Beavers as a relatively long-lived species produce 3–4 newborns every year. Were the young to stay with their parents and siblings, the colony would grow huge in a few years and soon outstrip its food resources. More importantly, grown-up offspring must find mates to start reproducing themselves. Mating with close relatives would often result in disastrous genetic defects, a phenomenon called *inbreeding depression.* For these reasons, beavers have to disperse from their native family.

Once beavers set out to disperse, they lose all benefits their home affords, including food and shelter. The path to their uncertain future is full of risks. For shelter, they have to make do with whatever presents itself on their journey. They first hide under roots or in bank holes. At ditches or small marginal streams, they may construct a token lodge, a "starter home" of sorts (Plate 29). Many large predators including wolves, coyotes, bears, and cougars lurk. Nowadays, highway traffic kills many traveling beavers. Most dispersers die before they make it to their final destination.

We must distinguish *natural dispersal* from *translocation* and *homing.* This chapter deals with natural dispersal, the process of leaving the parental colony and finding a site of one's own. If we remove beavers from one area and transplant them to another, they more often than not start roaming, sometimes inaccurately termed *homing.* They may move because they feel displaced into strange surroundings, but in many cases they do not travel in the direction of their original site. This second type of movement will be discussed in chapter 18.

Age

Usually, beavers leave their family at the age of 2 years, just before a new litter of siblings is born. However, they may delay their departure for 1 more year when the population density is high. Then, few suitable sites in the vicinity are vacant. If they stay at home, these 3-year-olds help their parents raise their younger brothers and sisters. In our study at Allegany State Park, we found that 64% of dispersers were 2 years old and 21% were 3 years old. Occasionally (14%), 1-year-olds also dispersed to a nearby colony if vacancies occurred.[1]

Timing

Dispersal is well timed. In the northern United States, beavers choose to disperse in spring right after the snow melts.[2,3] (For this reason, you may have noticed more beaver roadkills in spring than in any other season.) The thaw results in water flowing everywhere, easing the travel for beavers and permitting them to seek shelter when predators threaten. The many temporary streams and water-filled ditches enable beavers to reach many areas to settle in that in other seasons they would have to approach over land. With water all around, they can easily create a pond with minimal dam building at their chosen destination. In western Montana some beavers begin to disperse as late as August.[4]

The time of dispersal coincides with the time the new kits are born. We still do not know whether parents expel their young, possibly when the lodge becomes crowded, or the offspring leave on their own accord. Lodge space is not necessarily the issue, as many colonies already have more than one lodge and young beavers could build a new lodge on their home site. Also, delayed dispersal occurs frequently. But if parents expel 2-year-olds, how can the latter delay dispersal? Two-year-olds very likely decide by themselves whether to stay or to disperse, without the parents' interference. However, one of us (D.M.-S.) observed on one spring morning in the Huyck Preserve on New York's Helderberg Plateau, the parents of one lodge not permitting their whining yearlings to return to the lodge. As the sun rose and it became 10 and 11 o'clock, the two young were still floating in water-shield beds, with no better place to hide. Unfortunately, we do not know the precise circumstances of these interactions between adults and young.

Direction

Although mortality hits beavers hard during dispersal, they manage to counter this threat by effective strategies. First, they venture out on short excursions to assess their prospects of dispersal before they finally leave for good.[5] Some sallies extend over more than 2.5 km (1.5 miles) away from home. On these short trips beavers learn about "road conditions" for dispersal and whether vacant sites beckon in the neighborhood. They prefer to travel along streams. Water offers refuge from predators and keeps them cool, and bank holes come in handy for hiding. They try to minimize their time on land, primarily to avoid predators, even if shortcuts offer themselves.

Possibly to save energy, beavers head more often downstream (37%) than upstream (20%) when they start to disperse. However, since equal numbers of vacancies are available in both directions (assuming the stream does not peter out soon at nearby headwaters) on average, more beavers have to return empty-handed from downstream to try their luck upstream (17%) than the other way around (7%).[1]

Distances

Beavers do not travel farther than necessary. If suitable habitat is available near their native family, they will settle there. At Allegany State Park, we found that 35% of the dispersers had relocated at neighboring colonies. Males were more likely to colonize a neighboring site than females (31% vs. 17%). About 88% of the dispersers moved to a location within 5 km (3.1 miles) from their native colonies.[1] Short-distance dispersal obviously reduces the risks to beavers, but they have to pay a price. First, previous colonizers often have exhausted the desirable food resources in nearby habitats. So both quantity and quality of forage may leave much to be desired. As a result, the new settler has to make another move sooner or later after using up the remaining food. Therefore, long-distance dispersal in stages over several years is not uncommon.

Beavers can disperse more than 50 km (31 miles) along streams. Because beavers also move on land (although that is not their first choice), the actual dispersal distance may be shorter and less impressive. In our study, the longest distance traveled by beavers was 31.7 km (19.7 miles) as the crow flies. The animal may have traversed this distance in several dispersal stages over years, rather than in one shot. It may be significant that females in Allegany State Park dispersed farther (average: 10.15 km or 6.3 miles) than males (average: 3.5 km or 2.17 miles). Dispersers found beyond 20 km (12.4 miles) from their native colonies were all females.[1] Females need more and better food for successful breeding and possibly travel farther to find it. It is likely that in the small area near the colony of origin, the food has been severely depleted by years of use by beaver colonies. As the females increase their radius of exploration, the chance of finding a patch of little-used or recovered vegetation increases. Males, on the other hand, may be comparatively less dependent on quantity and quality of food. In any event, beavers represent one of the few exceptions among mammals where females disperse farther than males. The porcupine (*Erethizon epixanthum*) is another exception.

Table 12.1 lists some dispersal distances from other beaver studies.

Occasionally, beavers inherit a part of their parents' site and convert it into their own. We witnessed twice in Allegany State Park how families split in this way. In both cases a row of many impoundments along an extended stretch of stream constituted the parents' colony. The dispersers simply took over one or

Table 12.1 | Some Dispersal Distances of North American Beavers

Geographical Regions	Dispersal Distance	
	Mean	Maximum
New York[1]	A: Males—3.5 km (2.2) Females—10.2 km (6.3)	A: 31.7 km (19.7; a female)
Idaho[3]	A: 8.5 km (5.3) Males—8.3 km (5.2) Females—10.9 km (6.8)	A: 18.1 km (11.3; a male)
Minnesota[5]	A: 17.1 km (10.7) S: 29.9 km (18.7)	A: 49.6 km (31) S: 81.6 km (51)

Note: A, by air (straight line); S, along stream. Values in miles are given in parentheses.

two ponds at the far end, paired up with a newcomer, and established a territory of their own.

Duration

Beavers need weeks to months from the beginning of dispersal to when they finally settle down. A beaver that finds a suitable site immediately is really lucky. More often than not, beavers have to wander about before one becomes available. Some beavers roam for so long that they find a good site only late in the summer. One radio-collared disperser in Montana had been "floating" around for 181 days before settling down.[4] Such beavers have little time left before the harsh northern winter sets in. They have to promptly build all essential structures needed for survival such as a lodge, one or several dam(s), and a food cache in the water before the pond freezes over.

Between the onset of dispersal and their final destination, beavers may hunker down in temporary, suboptimal places. Some of our Allegany beavers eked out a living in culverts and tiny ditches. The open water surface can be as small as 2–3 m^2 (~20–30 square feet). However, these barely livable spots are very precious to their owners. They scent-mark all over their meager colony and tenaciously defend it against invaders. Other dispersing beavers may visit with relatives for several weeks en route to their uncertain destination. With the help of these relatives, dispersers will greatly avoid adverse conditions and succeed better in completing their dispersal. The dispersers and their unfamiliar relatives seem to recognize each other by odor from the anal gland secretion, by means of phenotype matching.

It appears that beavers usually find their mates during dispersal. "Floating" beavers we found in late spring and early summer often traveled in pairs. If a beaver lingers for an extended time, it may very likely bump into a desirable mate on its way to its future home. However, in Ohio pairs formed between a single resident who had lost its partner, or had founded a new colony, and a single immigrant attracted to the singleton.[6]

Even entire families may move to new areas, forced by a lack of water or food. This is quite different from the dispersal of young animals and should be well distinguished. Nor are the often considerable movements of captured and transplanted beavers true "dispersal."

REFERENCES

1. Sun, L., D. Müller-Schwarze, and B. A. Schulte. 2000. Dispersal pattern and effective population size of the beaver. Canadian Journal of Zoology 78: 393–398.
2. Leege, T. A., and R. M. Williams. 1967. Beaver productivity in Idaho. Journal of Wildlife Management 31: 326–332.
3. Leege, T. A. 1968. Natural movements of beavers in southeastern Idaho. Journal of Wildlife Management 32: 973–976.
4. Van Deelen, T. R., and D. H. Pletscher. 1996. Dispersal characteristics of two-year-old beavers, *Castor canadensis*, in western Montana. Canadian Field-Naturalist 110: 318–321.
5. Beer, J. R. 1955. Movements of tagged beaver. Journal of Wildlife Management 19: 492–493.
6. Svendsen, G. E. 1989. Pair formation, duration of pair-bonds, and mate replacement in a population of beavers (*Castor canadensis*). Canadian Journal of Zoology 67: 336–340.

Part IV

Ecology

chapter 13

Where They Live and Why: Habitat Requirements

> Sluggish streams and small lakes with clay banks that are well-wooded with aspen and willow are the favourite haunts of the Beaver. Streams that run in rocky beds, or that dry up in summer, and large rock-bound lakes are equally shunned.
>
> *E. T. Seton, 1909*

> The habitat of the American beaver is unusually broad. It is not surpassed by that of any other animal upon the continent, the deer and the fox not excepted. He was found from the confines of the Arctic Sea on the north, to the Gulf of Mexico, the Rio Grande, and the Gila rivers on the south, and southward of these ranges in Tamaulipas in Mexico, which is the southernmost point to which he has been definitely traced.
>
> *Lewis H. Morgan, 1868*

Beaver, Beaver, Everywhere?

Before Europeans arrived in North America, beavers used to range across the entire United States (except Florida and a few desert areas) and a large part of Canada (except the Arctic tundra). Today they dwell from the subarctic to the Rio Grande, which separates the United States from Mexico. Beavers colonize elevations as high as the timberline in Colorado.[1] In the eastern United States we found beavers in extreme places. At the base of Mt. Marcy (to Indians *Tahawus*, the "Cloud Splitter"), the highest Adirondack mountain, lies Lake Tear of the Clouds, which spawns the mighty Hudson River. The lake owes its existence to a beaver dam. We found beavers even here (Plate 30), getting by on paper birch. Balsam fir dominates around Lake Tear of the Clouds, so beavers have to search for the occasional birch tree. But the lake provides a reliable water supply. Better yet, at 4000 feet (1219 m) on Mt. Marshall, a trailless peak in the Adirondacks, beavers are eking out an existence in a tiny pond, getting by on balsam fir, the only tree present. Balsam fir boasts rich resin reservoirs, once used by Indians for sealing containers and canoes, truly not a first choice for ordinary beavers. This beaver site has existed for a long time, as it is mentioned in the hiking guide for the high peaks of the Adirondacks. On Mt. St. Helens in Washington State, we once again see beavers, and their signs and structures in numerous places, 2 decades after the enormous eruption of the volcano in May 1980. Beavers have reclaimed areas as close to the volcano as Coldwater Lake and Meta Lake. They probably have also ventured into Spirit Lake, a lifeless zone immediately after the mountain blew its top.

To be sure, these extremes are the exceptions to the rule. If given a choice, beavers prefer less steep terrain and diverse vegetation. Plates 30 to 34 show a variety of habitats beavers have colonized. The Alaskan landscape at Denali National Park (Plate 31) lacks trees; beavers use willow brush. Beavers near the Continental Divide in Rocky Mountain National Park, Colorado, have ample cover in luxuriant willow stands along streams (Plate 32). In the Zuni Mountains, New Mexico, beavers have to seek out the few cottonwood and aspen trees along streams in an otherwise solid pine and juniper forest (Plate 33). At the southern end of the Western Hemisphere, introduced beavers have taken well to an entirely new diet of southern beech (*Nothofagus* sp.). Finally, developed landscapes, such as we find in most of Europe, pose no obstacle for beaver colonization (Plate 34).

Water

First and foremost, beavers need water. Without water, the best food supply will not attract them. Water can be in the form of a stream, river, lake, or pond, as long as there is a year-round supply sufficient for swimming, diving, floating logs, protection of lodge and burrow entrances, and general safety from predators. Beavers always escape into the water; they are extremely wary on land. On streams, beavers prefer slowly moving water. Typical stream gradients are 1% and below; in one study, the gradients at beaver sites averaged 1.16%.[2] In eastern Oregon, an arid area, gradients averaged 2.3%.[3] However, under conditions of high population density, beavers may utilize streams with even steeper gradients, up to 6.5%, as we found in Allegany State Park in western New York State. In mountainous Colorado most beavers colonize streams with a gradient of less than 6%; 15% may be the limit.[1] Wider streams (e.g., 8 m wide) are preferred over narrower ones (about 1.4 m wide).[3]

The Eurasian beaver in the Rhone Valley of France, as in other parts of Europe, prefers streams with gentle, uniform flow and sufficient water depth. In addition, beavers require the presence of willow bushes.[4]

Food

Beavers need a reliable food supply but are quite flexible in this regard. They are generalists, always consuming a variety of plant species. While a reliable water supply is a must, beavers can manage in many different types of vegetation. This great variety of acceptable vegetation types ranges from aspen groves to cornfields. However, they need woody vegetation for the winter. Therefore, in the north beaver range as far as willows will grow. Trees and shrubs also serve as building materials for dams and lodges. Lacking such construction wood, beavers are limited to living in bank holes. The types of trees and their size (age) and distance from the water's edge all affect where beavers choose to settle.

North American beavers prefer aspen (genus *Populus*), and the beaver's geographical distribution coincides almost completely with that of aspen. In Europe, beavers are partial to willow. For instance, in Austria, Eurasian beavers living side by side with introduced North American beavers prefer willow, even in mixed forests, while their New World cousins cut all species present (J. Sieber, personal communication, 1989). Therefore, the most preferred habitat is where aspen or willows abound. These trees grow fast and have soft wood that is easy to chop down and peel. Beavers will colonize an area where aspen is available within a reasonable distance from the water, about 60 m, but beavers will harvest aspen as far away from their pond as several hundred meters.[5] After forest fires that destroy mature stands, aspen often appears as a first successional species. Such bursts of acres and acres of aspen saplings invite beavers to settle and lead to population explosions. In the Adirondack Mountains, such fires occurred in the early 20th century, especially in 1903 and 1909.

A slowly meandering stream with aspen trees and alder or willow thickets near the water compromises the ideal beaver habitat. Beavers can also thrive in hardwood stands that are composed of maples, alders, cherry, beech, and hornbeam.[6,7] If the food supply is exhausted after several years of harvesting, the entire family will move to an area with a better food supply. We have seen several cases of beaver movements at Allegany State Park in New York. As beaver populations grow denser, and vegetation becomes depleted, beavers resort to less preferred tree species. In New York State, after aspen and willow, their order of preference ranges from birch, black cherry, beech, juneberry, and hornbeam to maples, hawthorn, and hemlock. The least-preferred species are the conifers such as balsam fir, white pine, Scots pine, red pine, and Norway spruce. Conifers often produce high levels of resins that defend against herbivores, including insects, birds, and mammals. In late winter, when nutritionally stressed, and the sap starts to flow in trees, beavers sometimes eat the bark of pines (See Plate 22), while spruce is rarely taken.[7] When beavers have depleted hardwoods and start to consume pines, they might leave the site after several weeks. Therefore, a good number of felled and consumed conifer trees signify that a beaver family might move.

The nonwoody vegetation eaten during the growing season includes aquatic plants such as water shield, or water and pond lilies, and ferns, raspberry, grasses, and sedges. In Nebraska, beavers feed on corn and store dry ears of corn as winter food. Similarly, in Bavaria beavers consume sugar beets, corn, and other grains.

In Ohio, beavers resided on average about 4 years at the same site before they moved on.[8] After food plants, especially trees, grow back, beavers may return and reclaim the same site. In the Colorado Rocky Mountains, beaver families have recolonized the same site repeatedly for 70 years! While present at a site, a beaver family may occasionally shift its center of activity and use different, adjacent ponds in subsequent years. This provides depleted vegetation near ponds in disuse time to recover.

Suitable food resources should grow within 30 m from water colonized by beavers.[9,10] Beavers are awkward and slow on land, exposing themselves to predators such as wolves, bears, and coyotes. Hence, beavers minimize their time on land and stay close to the water. This permits them to jump into the water at the slightest potential danger. Logging trees far from the water not only costs time and energy (to haul tree parts to the water to process them there) but also increases the risk of being preyed on on land.

For beavers living in lakes, the water level is relatively stable over time. Such "lake beavers" do not have to spend time and energy to build dams. They usually build a lodge at or near the bank where the water is shallow. For these beavers, available food determines where they settle. At Cranberry Lake in the Adirondack Mountains in New York, beavers sometimes have to swim as far as 1–2 km (a mile, more or less) from their lodge to a suitable food source. There we observed a beaver colony in the summer of 1995. The beavers left their lodge for feeding places at dusk and did not return until the morning of the next day. Some beavers swam 1–2 km (1 mile) along the shoreline of the lake. Finally they landed and chopped down trees right behind our cabins while we were sleeping soundly!

Temporary Habitat

Dispersers and beaver families may occupy temporary homes on the way to their final destinations. This happens most frequently in spring. Then you can see beaver signs, such as tracks in mud, tracks in snow, and droppings in the water in unexpected places. Ditches, small streams, temporary swamps, even your backyard may show signs of feeding by beavers. However, the animals usually will not stay there for long because these "marginal sites" lack the essential resources for subsistence. For instance, the water may dry out intermittently.

Coping with Flooding

Beavers normally avoid places that flood often. In the fall of 1996, we found a series of about 20 dams belonging to four families near the mouth of Umtanum Creek, which feeds the Yakima River at Ellensburg, Washington. After an unusually high level of flooding in the spring of 1997, no beavers resided there for the following 4 years.

However, beavers do not depend only on what is naturally available to them. They enjoy a unique ability to actively modify the landscape to create a habitat better suited for their needs.[10] At Allegany State Park, beavers living in Red House Creek often suffer from inundation during the summer, when sudden thunderstorms or torrential rains flood the land. Although beaver dams are engineering marvels that last several years by virtue of maintenance and continuous strengthening, many dams still become seriously damaged or totally washed out. In 1 year we lost two-thirds of 20 study colonies this way. The ponds drained completely. After the flood, the beavers came back to reclaim their home and quickly rebuilt all structures, including the damaged lodge and dams. Really successful beavers prevent such catastrophes by building a small pond off the main course of a stream, branching off only some water, like the millrace of old. This "design" has a disadvantage: at low stream levels, fresh inflow may stop, and the water in the pond becomes stagnant or even dries out intermittently.

Developed Landscape

As beaver populations grow, they recolonize former beaver habitat that is now occupied or used by humans. In North America, beavers move into suburbs, golf courses, even shopping centers, leading to conflicts with people, as we discuss in detail in chapter 19. Such encounters with the developed landscape are the norm in Europe. In Bavaria, beavers settle along ditches with only narrow strips of woody vegetation that separate the watercourse from agricultural fields. The animals adapt to the scarcity of trees by utilizing a long stretch of the ditch or stream. In summer, six measured home ranges ranged from 1770 to 3270 linear meters, while those distances shrank to 300–820 m in winter.[11] Flexible as beavers are, they compensate for a lack of natural vegetation by foraging in agricultural fields. Among the 300 documented plant species eaten, they consume corn (maize), other grains, and sugar beets.[11] Plate 34 shows a bank lodge at a backwater in the floodplain of the River Elbe in central Germany. Note the narrow strip of trees between the watercourse and the pastures and fields beyond. In keeping with the flexibility of the Eurasian beaver, human activity in the highly developed Rhone Valley in France did not seem to keep beavers from an otherwise suitable habitat.[4] As another example, beavers in the Czech Republic coexist with human use of their habitat (V. Kostkan, personal communication, 2000). In Slovakia, beavers immigrated from Austria. They even have taken up residence in Greater Bratislava, the country's capital, although two-thirds of the animals stayed at their sites for only 1 or 2 years.[12]

In summary then, both beaver species prefer habitats that provide a secure water supply and choice plants. But they manage to adapt to a variety of landscapes, even those permanently and thoroughly modified by humans.

REFERENCES

1. Retzer, J. L., H. M. Swope, J. D. Remington, and W. H. Rutherford. 1956. Suitability of physical factors for beaver management in the Rocky Mountains of Colorado. Technical Bulletin 2. Denver: Colorado Department of Game and Fish.
2. Beier, P., and R. H. Barrett. 1987. Beaver habitat use and impact in Truckee River Basin, California. Journal of Wildlife Management 51: 794–799.
3. McComb, W. C., J. R. Sedell, and T. D. Buchholz. 1990. Dam-selection by beavers in an eastern Oregon Basin. Great Basin Naturalist 50: 273–281.
4. Erome, G. 1982. Contribution a la connaissance Éco-Éthologique du castor (*Castor fiber*) dans la vallée du Rhône [doctoral dissertation]. Université Claude Bernard, Lyon, France.
5. Müller-Schwarze, D., and B. A. Schulte. 1999. Behavioral and ecological characteristics of a "climax" population of beaver (*Castor canadensis*). In: P. E. Busher and R. M. Dzięciołowski, editors. Beaver protection, management, and utilization in Europe and North America. New York: Kluwer Academic/Plenum. p 161–177.
6. Hammond, M. C. 1943. Beaver on the Lower Souris refuge. Journal of Wildlife Management 7: 316–321.
7. Müller-Schwarze, D., B. A. Schulte, L. Sun, A. Müller-Schwarze, and C. Müller-Schwarze. 1994. Red maple (*Acer rubrum*) inhibits feeding by beaver (*Castor canadensis*). Journal of Chemical Ecology 20: 2021–2034.
8. Svendsen, G. E. 1989. Pair formation, duration of pair-bonds, and mate replacement in a population of beavers (*Castor canadensis*). Canadian Journal of Zoology 67: 336–340.
9. Belovsky, G. E. 1984. Summer diet optimization by beaver. American Midland Naturalist 111: 209–222.
10. Liedholdt, K. L., W. McComb, and D. E. Hibbs. 1989. The effect of beaver on stream and stream-side characteristics and coho populations in western Oregon. Northwest Science 63: 71.
11. Schwab, G., W. Dietzen, and G. V. Lossow. 1994. Biber in Bayern—Entwicklung eines Gesamtkonzeptes zum Schutz des Bibers. Beiträge zum Artenschutz 18: 9–44. Munich: Bayerisches Landesamt fuer Umweltschutz.
12. Pachinger, K., and T. Hulik. 1999. Beavers in an urban landscape. In: P. E. Busher and R. M. Dzięciołowski, editors. Beaver protection, management, and utilization in Europe and North America. New York: Kluwer Academic/Plenum. p 53–60.

chapter 14 Mortality and Predators

> I came upon a beaver house that was surrounded by a pack of wolves. These beasts were trying to break into the house. Apparently an early autumn snow had blanketed the house and thus prevented its walls from freezing. The soft condition of the walls, along with the extreme hunger of the wolves, led to this assault. Two of these animals were near the top of the house clawing away at a rapid rate. Now and then one of the sticks or poles in the house-wall was encountered, and at this the wolf would bite and tear furiously. . . . A number of wolves lay about expectant; a few sat up eagerly on haunches, while others moved about snarling, driving the others off a few yards, to be in turn driven off themselves. Shortly before they discovered me, there was a fierce fight on top of the house, in which several mixed.
>
> *Enos A. Mills, 1913*

Mortality

Without mortality at young stages, beaver populations would grow rapidly to enormous numbers, as pointed out in chapter 11. The total annual mortality of beavers of all age classes was 30% in one population in Newfoundland,[1] and 27% in another study in Newfoundland.[2] During the first 6 months of life, as many as 52% may perish.[2] In the first 2–3 years of life, mortality usually is highest, and between the ages of 5 and 9 years very few beavers die in unexploited populations.[3] In most beaver populations the total mortality consists of two components: natural deaths from predators, accidents, disease, starvation, and so on, and deaths from trapping. In Newfoundland, Payne[2] found natural mortality to be 0%–24% and harvest mortality to be 11%–36%. Mortality starts in utero: prenatal mortality by resorption of embryos can reduce the number of young by 11%–16%.[3]

Mortality varies geographically. As many as 30% of all beavers may die during a severe Newfoundland winter. Mortality is particularly high in young beavers, as 46% of kits, yearlings, and 2-year-olds but only 12% of the adults may perish.[1] In Massachusetts, fewer, but still 28%, of the kits might vanish.[4]

Predators

Owing to their impenetrable fortress, the often free-standing lodge with a "moat" for security, and their reluctance to venture far from the safety of water, beavers have few predators. Today, humans constitute the most important predator that regulates beaver densities, by fur trapping or by removal.

In the natural state, wolves (*Canis lupus*) and coyotes (*C. latrans*) can take significant numbers of beavers, if other prey such as deer become scarce.[5–7] Beaver remains occurred in 7% of wolf scats in the Rocky Mountain national parks of Canada[8] and in 10.5% in Ontario.[9] On Isle Royale in Lake Superior, 11% of wolf scats collected during 3 years contained beaver remains.[5] However, in the years 1952–1955 wolf scats with beaver remains comprised 35% of the total found on the island.[5] In Algonquin Provincial Park, Ontario, Canada, up to 62.8% of wolf scats contained beaver hair in summer, but only 11.5% did in winter.[10] A somewhat different seasonal pattern was reported from Minnesota: 20%–47% of wolf scats found in April and May contained beaver remains, but only 2% found in June and July contained them.[11] In southern Quebec beavers constituted 29%–44% of the biomass in the diet of wolves from May through November but only 1%–3% during the remainder of the year.[12] Table 14.1 summarizes these findings.

Experimental wolf reduction in the province of Quebec suggested an important impact of wolves on beavers: beaver numbers increased when managers reduced the wolf numbers, but decreased when the wolf population was allowed to grow again.[13] It is thought that wolf predation caused the decline of the beaver population on Isle Royale in the 1980s.[14] Beavers are vulnerable to wolves only during fall and early spring when they venture farther from their pond in search of food. In winter when wolves can travel easily on ice, the beavers remain safely in their lodges. In northern Manitoba, wolves have been observed to stalk beavers.[15] According to the prevailing view of researchers, wolves turn to beavers as prey only when larger prey such as moose and deer decline in numbers.

Coyotes have been observed to hunt and kill beavers in Rocky Mountain National Park, Colorado.[16] In the Adirondack Mountains of New York, beavers are

Table 14.1 | The Wolf as Beaver Predator: Beaver Remains in Wolf Scats

Area	Percentage of Wolf Scats with Beaver Remains
Canadian national parks[8]	7
Ontario[9]	10.5
Isle Royale at Lake Superior[5]	11
Minnesota[11]	20–47 in April/May
	2 in June/July
Algonquin Provincial Park, Ontario, Canada[10]	62.8 in summer
	11.5 in winter

the second most important prey species of coyotes, after white-tailed deer. In spring, 10.1% of coyote scats contained beaver remains, and summer values were also high (9.2 and 8.7%, respectively). In winter, when beavers stay in their lodges most of the time, fewer coyote scats (5.6%) included beaver remains.[17]

Other potential predators include black bear (*Ursus americanus*), river otters (*Lutra* sp.), red wolf (*Canis rufus*), and cougar (*Felis concolor*). Members of our group have observed black bears taking beavers out of live traps on at least two occasions.[18,19] In northern Saskatchewan it was "fairly common" for black bears to dig into beaver lodges.[3] Significant predation by black bears on beavers has only been reported from an island in Lake Superior.[20] According to Murie,[21] grizzly bears (*U. arctos*) in Alaska provided "no observations suggesting bear predation on beavers." Beavers most likely offer only an "occasional taste of carrion for the bears."

Fox (*Vulpes vulpes*), bobcat (*Lynx rufus*), and lynx (*L. canadensis*) prey little on beavers.[22] In Maine, less than 5% of bobcat carcasses contained remains of beavers in their digestive tract.[23] For the obvious reason that both prey and predators operate mostly during darkness, attacks on beavers have seldom been witnessed. One such case was observed in Norway. A red fox attacked and killed a kit of the Eurasian beaver 15 m from the lodge, dragged it 10 m to a tree, and consumed it. Ten minutes later an adult swam toward the scene and appeared to challenge the fox. It slapped its tail and finally dived and disappeared. Having scared the fox away, the observer examined the carcass. The fox had opened the chest and eaten the sternum, heart, lungs, stomach, cecum, both intestines, and most of the liver and ribs and some forearm muscle. Thirteen vertebrae were missing. The wounds suggested that the fox had bitten between the shoulders and presumably severed the spinal cord, resulting in a quick death.[24]

Despite claims to the contrary, the river otter (*Lutra canadensis*) is not a predator of beavers. Beaver remains are seldom found in otter scats. In an area with high beaver and otter densities in Alberta, Canada, Reid[25] found only 5 otter scats (0.4%) with beaver remains out of 1140, possibly obtained from beaver carcasses. No beaver remains were present in otter droppings in seven other studies cited by Reid.[25]

Since alligators (*Alligator mississippiensis*) are known to prey on nutria (*Myocastor coypus*), Arner and coworkers[26] investigated whether alligators kill beavers and whether they could be used to control "nuisance beaver" populations. Four large alligators and 6 adult beavers were released in a pond in Mississippi. Five of the 6 beavers were killed and eaten by the alligators. For possible control of beaver numbers, 33 alligators were released in an area of beaver impoundments that had created a 400-ha hardwood swamp. Beaver activity in the area and trapping results indicated that the alligators had reduced the beaver population, notably the younger age classes: in the alligator area no yearlings were trapped, while in a con-

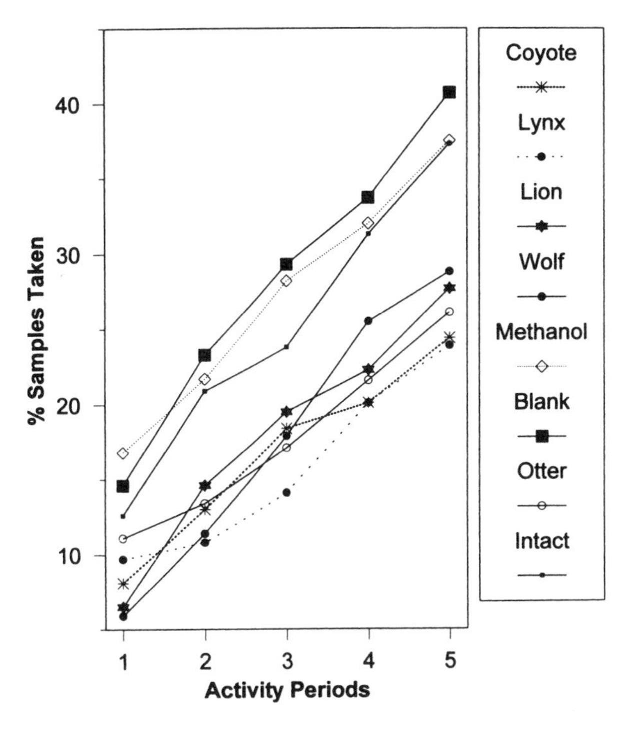

Figure 14.1 | Predator odor experiment. Extracts from predator droppings applied to aspen sticks affect feeding by beavers. They avoided eating bark from sticks treated with extracts from lynx, coyote, otter, and lion (lower group of curves). The most accepted samples were 3 controls (upper 3 curves): completely untreated (Intact), punctured (Blank), and treated with solvent only (Methanol). Curves show cumulative removal of sticks over 5 nights (activity periods) of the experiment. (From A. Engelhart and D. Müller-Schwarze. 1995. Responses of beaver [*Castor canadensis* Kuhl] to predator chemicals. Journal of Chemical Ecology 21: 1357, Figure 4. Reproduced with permission from Kluwer Academic/Plenum.)

trol area 5 of 14 trapped beavers were yearlings. The researchers concluded that 2-m-long alligators can kill beavers of up to 13 kg body weight. But during the hot summer months, the alligators moved from the shallow beaver ponds to deeper ponds and streams. This behavior reduces the usefulness of alligators for beaver control.

Antipredator Behavior

In field experiments, beavers in both North America[19] and Europe[27] avoided food with predator odors. The results were remarkably similar for both beaver species. Aspen sticks treated with extracts from feces of wolf, red fox, dog, river otter, lynx, lion, and black and brown bears were eaten less often than untreated control sticks. But the beavers did not significantly discriminate between the various predator chemicals. Still, in North America lynx and coyote odors were slightly more effective than the others (Fig. 14.1), and in Europe red fox and river otter odors topped the list as repellents.

Other Causes of Mortality

Trapping or shooting is the most significant cause of mortality. Roadkills are common, especially in spring. In Europe, one-third to one-half of dead beavers died of diseases. Starvation in hard winters is another factor. On rare occasions, beavers have been killed when felling trees. Nitsche[28] reviewed the reported incidents in the Eurasian beaver (*C. fiber*).

REFERENCES

1. Bergerud, A. T., and D. R. Miller. 1976. Population dynamics of Newfoundland beaver. Canadian Journal of Zoology 55: 1480–1492.
2. Payne, N. F. 1984. Mortality rates of beaver in Newfoundland. Journal of Wildlife Management 48: 117–126.
3. Gunson, J. R. 1970. Dynamics of the beaver of Saskatchewan's northern forest [M.S. thesis]. Edmonton: University of Alberta.
4. Brooks, R. P., M. W. Fleming, and J. J. Kenelly. 1980. Beaver colony response to fertility control: evaluating a concept. Journal of Wildlife Management 44: 568–575.
5. Mech, L. D. 1966. The wolves of Isle Royale. U.S. Park Service, Fauna Series 7. Washington, D.C.: U.S. Government Printing Office.
6. Pimlott, D. H., J. A. Shannon, and G. B. Kolenosky. 1969. The ecology of the timber wolf in Algonquin Provincial Park. Toronto: Ontario Department of Land and Forests.
7. Allen, D. L. 1979. Wolves of Minong: their vital role in a wild community. Ann Arbor: University of Michigan Press.
8. Cowan, I. M. 1947. The timber wolf in the Rocky Mountain national parks of Canada. Canadian Journal of Research 25: 139–174.
9. Peterson, R. L. 1955. North American moose. Toronto: University of Toronto Press.
10. Theberge, J. B., S. M. Oosenburg, and D. H. Pimlott. 1978. Site and seasonal variations in food of wolves in Algonquin Park, Ontario. Canadian Field-Naturalist 92: 91–94.
11. Fuller, T. K. 1989. Population dynamics of wolves in north-central Minnesota. Wildlife Monograph 105: 1–41.
12. Potvin, F., H. Jolicoeur, and J. Huot. 1988. Wolf diet and prey selectivity during two periods for deer in Québec: decline versus expansion. Canadian Journal of Zoology 66: 1274–1279.
13. Potvin, F., L. Breton, C. Pilon, and M. Macquart. 1992. Impact of experimental wolf reduction on beaver in Papineau-Labelle Reserve, Quebec. Canadian Journal of Zoology 70: 180–183.
14. Shelton, P. C. and R. O. Peterson. 1983. Beaver, wolf, and moose interactions in Isle Royale National Park, USA. Acta Zoologica Fennica 174: 265–266.
15. Nash, J. B. 1951. An investigation of some problems of ecology of the beaver *Castor canadensis* Kuhl in northern Manitoba [M. Sc. thesis]. Winnipeg, Canada: University of Manitoba.
16. Packard, F. M. 1940. Beaver killed by coyotes. Journal of Mammalogy 21: 359–360.

17. Brundage, G. C. 1993. Predation ecology of the eastern coyote, *Canis latrans* var., in the Adirondacks, New York. [M. Sc. thesis]. Syracuse: State University of New York, College of Environmental Science and Forestry.
18. Schulte, B. A. 1993. Chemical communication and ecology of the North American beaver (*Castor canadensis*) [Ph.D. dissertation]. Syracuse: State University of New York College of Environmental Science and Forestry.
19. Engelhart, A., and D. Müller-Schwarze. 1995. Responses of beaver (*Castor canadensis* KUHL) to predator chemicals. Journal of Chemical Ecology 21: 1349–1364.
20. Smith, D. W., D. R. Trauba, R. K. Anderson, and R. O. Peterson. 1994. Black bear predation on beavers on an island in Lake Superior. American Midland Naturalist 132: 248–255.
21. Murie, A. 1981. The grizzlies of Mt. McKinley. Washington, D.C.: U.S. Department of the Interior, National Park Service.
22. Saunders, D. A. 1988. Adirondack mammals. Syracuse: State University of New York College of Environmental Science and Forestry.
23. Litvaitis, J. A., A. G. Clark, and J. J. Hunt. 1986. Prey selection and fat deposits of bobcats (*Felis rufus*) during autumn and winter in Maine. Journal of Mammalogy 67: 389–392.
24. Kile, N. B., P. J. Nakken, F. Rosell, and S. Espeland. 1996. Red fox, *Vulpes vulpes*, kills a European beaver, *Castor fiber*, kit. Canadian Field-Naturalist 110: 338–339.
25. Reid, D. G. 1984. Ecological interactions of river otter and beavers in a boreal ecosystem [M.S. thesis]. Calgary: University of Calgary.
26. Arner, D. H., C. Mason, and C. J. Perkins. 1981. Practicality of reducing a beaver population through the release of alligators. In: J. A. Chapman and D. Pursley, editors. Proceedings of Worldwide Furbearer Conference; 1980 Aug. 3–11; Frostburg, Md. Volume 3. Frostburg, Md.: Worldwide Furbearer Conference, Inc. p 1799–1805.
27. Rosell, F., and A. Czech. 2000. Responses of foraging Eurasian beavers *Castor fiber* to predator odours. Wildlife Biology 6: 13–21.
28. Nitsche, K. A. 1985. Unfälle des Bibers beim Baumfällen. Mitteilungen der Zoologischen Gesellschaft Braunau 4: 275–276.

chapter 15 Parasites and Diseases

Some kind of distemper was prevailing among the animals which destroyed them in large numbers. I found them dead and dying in the water, on the ice, and on the land. . . . Many of them which I opened were red and bloody about the heart. Those in large rivers and running water suffered less; almost all of those that lived in ponds and stagnant water died.

John Tanner, ca. 1800

Upon opening carcasses the only abnormal condition reported by the early trappers was a bloody congestion about the heart. The cause of these early die-offs will, of course, never be known, but in light of recent events tularemia may logically be suspected. . . . The field evidence is certainly suggestive that [the tick] *Ixodes banksi* may be involved in the transmission of tularemia.

W. H. Lawrence, L. D. Fay, and S. A. Graham, 1956

Two human diseases linked to endoparasites of beavers have repeatedly made headlines: tularemia and *Giardiasis.* In addition to such endoparasites, beavers may be host to a number of helminth (worm) species. Tuberculosis epidemics have also been described for *C. fiber* in the wild and *C. canadensis* in captivity. Ectoparasites, on the other hand, include beetles and mites. Finally, tumors occur occasionally (Fig. 15.1).

Endoparasites

Tularemia is caused by the bacterium *Pasteurella tularensis,* named after Tulare County in California. The bacteria are found in blood, all organs, body fluids, and excreta. For instance, in 1981 and 1982, beaver mortality due to tularemia was "substantial" at the Necedah National Wildlife Refuge and adjacent state wilderness areas in central Wisconsin.[1] The bacteria contaminate the water when infected beavers die in the water. Humans contract the disease by handling infected

Figure 15.1 | A beaver with a large tumor. Allegany State Park.

carcasses and consuming contaminated water. Bites by mosquitoes and ticks that have fed on infected beavers also transmit the disease. Because of the danger to people, the affected areas in Wisconsin were closed to the public for the spring and early summer of both years. Economically, tularemia becomes locally significant by reducing recreational opportunities. The North American beaver is more susceptible to tularemia than its Eurasian counterpart. In 1960, 116 of 150 beavers in a farm in Oregon succumbed to this disease.[2] During an outbreak in Voronezh Province, Russia, only 50% of *C. canadensis* animals but all of the local species, even when living together with diseased North American beavers, survived.[3]

Pseudotuberculosis (rodentiosis), caused by *Bacterium pseudotuberculosis rodentium*, is widespread among rodents, and can occur in beavers. Rabies is rare in beavers. Nevertheless, a case was reported in Florida in early 2002.

Human Health: "Beaver Fever"

Beavers have been implicated as a link in the chain that leads to an infection with the protozoan *G. lamblia* in humans,[4] first seen as early as 1681 by Leeuwenhoek in his own stool (Fig. 15.2). The infection, giardiasis, is popularly known as "beaver fever."

Campers who drink stream water or residents who receive their drinking water from reservoirs that are frequented by beavers may contract beaver fever. In 1984,

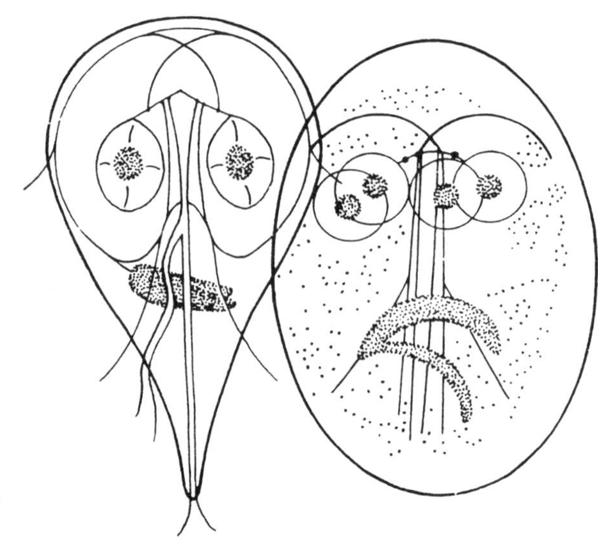

Figure 15.2 | A trophocyte (vegetative form) (left) and cyst of *Giardia lamblia* (right). (From N. D. Levine. 1979. *Giardia lamblia:* classification, structure, identification. In: W. Jakubowski and J. C. Hoff, editors. Waterborne transmission of giardiasis. Washington, D.C.: U.S. Environmental Protection Agency. p 2–8.)

for instance, 110,000 inhabitants of 17 towns in northeastern Pennsylvania had to boil their water because 63 people had contracted giardiasis. The 100 or so beavers living in the watershed that feeds the Springbrook Reservoir near Wilkes-Barre were implicated, but the issue remained controversial.[5]

The symptoms of this disease are diarrhea, abdominal cramps, gassiness, and weight loss. Vomiting, chills, headache, and fever characterize more severe cases. The symptoms appear about 2 weeks after infection with *G. lamblia.* Prescription drugs such as quinacrine (Atabrine) are used for treatment. The symptoms continue for about 10–15 days after the start of treatment. Streams become contaminated by untreated human sewage, which can contain 333–2000 *Giardia* cysts/gallon. Aquatic mammals such as beavers and muskrats contract giardiasis, concentrate the organism in their intestines, and pass large numbers of cysts with their feces into bodies of surface water.

Cysts in the water can be found microscopically by inspecting samples that have been concentrated by filtering. The cysts are slightly larger than a red blood cell. In a Pennsylvania study, 10% of commercially trapped beavers carried *Giardia* cysts.[6] Humans infect themselves with these cysts by drinking untreated water. Only a few cysts suffice to infect a person, as the organism multiplies rapidly. After a human has ingested the cyst, the parasite leaves the cyst and starts to multiply. These vegetative forms attach to the cells lining the intestinal wall. They feed on mucous secretions and block the absorption of fats and other nutrients from the intestine. This leads to fatty, foul-smelling diarrhea that can be explosive and watery. A single stool can contain up to 300 million cysts![7] Within families, the disease is contagious through interpersonal contact.

Between 1965 and 1981, 53 outbreaks were reported in the United States, affecting over 20,000 people. *Giardia* cysts are resistant to the levels of chlorination in drinking water that inactivate coliform bacteria. Levels that destroy the cysts render the water unpalatable to humans.[8] When camping, one should take precautions such as boiling, filtration, or chemical treatment, when surface water must be used.

It needs to be emphasized that the primary source of a *Giardia* outbreak is contamination of surface waters by humans. The beaver can only concentrate the parasite put in the water by humans. In that sense, the beaver is not the source but merely a carrier of the parasite. In recreational areas, the infections often come and go with the vacation season. Beavers lose the *Giardia* cysts over the winter and become reinfected in summer. In Fraser, Colorado, beavers upstream of a sewage plant tested negative for *Giardia* but were carriers below the plant.[4] More information can be found in the book edited by Erlandsen and Meyer.[8]

Helminths (worms) have been found in both beaver species. The trematode *Stichorchis subtriquetrus* occurs in the caecum of both Old and New World beavers. The liver fluke *Fasciola hepatica* has also been found in *C. fiber*. Of two prevalent nematodes, *Castorstrongylus castoris* lives in the large intestine of *C. canadensis* and *Travassosius americanus* and *T. rufus* in the stomach and small intestine of New and Old World beavers, respectively. In a study in Alberta, *T. americanus* and *S. subtriquetrus* were common and abundant. They occurred in 93% and 72% of the beavers, respectively.[9] Likewise, in Eurasian beavers, the most common helminths are *T. rufus* and *S. subtriquetrus*, found in 41% and 86% of the individuals, respectively.[10,11] Thus, the genera *Travassosius* and *Stichorchis* constitute the principal helminth fauna of beaver of the Holarctic. They have never been found in other animals. Including the less common species, 11 helminths have been identified in *C. canadensis* and 21, in *C. fiber*.[10] Many of these are restricted geographically or have other hosts, such as muskrats, and mallards.

Ectoparasites

Beetles

The beaver beetle (*Platypsyllus castoris*) belongs to Sylphidae (carrion beetles). Lacking hindwings, it can move from beaver to beaver only during direct body contact or in the lodge. The gut of beetles found on Eurasian beavers at the River Elbe contained beaver hair and epidermal skin fragments.[12] Both beaver species harbor this flightless and blind parasite. These features have spawned debates on whether the beetle—and also the beaver mite—have ridden along as the beaver spread from Eurasia over the Beringia Bridge to North America, or whether transplanted *C. canadensis* infected the Eurasian beaver. The latter is held unlikely today.

A second, North American beaver beetle is *Leptinillus validus.* The beetle and its larva appear to feed on mites, found on the head of the beaver, especially around the oral angles and the ears. The mite in turn most likely eats sebaceous secretion on the skin and hair.

Mites

Beaver mites all belong to the genus *Schizocarpus.*[13,14] Of these, only the original *S. mingaudi* occurs on both the Eurasian and the North American beaver. At least 15 species occur on the North American beaver and at least 33 on the Eurasian beaver. Within North America, the mite communities on beavers vary considerably: all the most abundant species on Alaskan beavers were different from the most abundant species on beavers from Indiana.[15] One beaver may carry about 10 species. As many as 133,000 mites may reside on a 54–pound (24.5-kg) beaver (J. Whitaker, personal communication). North American mites form four groups, which differ by the sucker plates on the males (females are very similar and cannot be distinguished within groups).[15] The different groups live on different parts of the beaver: 4 related species on the head and nape, a group of 6 other species on the back and outside of the hindlegs, a third group of 6 species on the belly, and 1 isolated species over the entire body of the North American beaver. Such a proliferation of species on the body of one host has been termed "multispeciation."[15]

REFERENCES

1. U.S. Department of Interior Fish and Wildlife Service. 1982. In: Reports on Fisheries and Wildlife Research. Denver, Colo.: Fish and Wildlife Service. p 105.
2. Bell et al., 1962, cited in Lavrov, L. S. 1983. Evolutionary development of the genus *Castor* and taxonomy of the contemporary beavers of Eurasia. Acta Zoologica Fennica 174: 87–90.
3. Avrorov and Borisov, 1947, cited in Lavrov, L. S. 1983. Evolutionary development of the genus *Castor* and taxonomy of the contemporary beavers of Eurasia. Acta Zoologica Fennica 174: 87–90.
4. Levine, N. D. 1979. *Giardia lamblia*: classification, structure, identification. In: W. Jakubowski and J. C. Hoff, editors. Waterborne transmission of giardiasis. Washington, D.C.: U.S. Environmental Protection Agency. p 2–8.
5. Anonymous. 1984. Pennsylvanians' illness may be caused by beavers. New York Times, Jan. 3, 1984. p B16.
6. Boutros, S. N., and O. W. Boutros. 1984. Survey of commercially trapped beaver *Castor canadensis* in Pennsylvania for *Giardia* sp., 1982 trapping season [abstract]. Presented at the annual meeting of American Society of Parasitologists, August, 5–9; Snowbird, Utah.

7. Ferris, M. 1986. Update on giardiasis. Adirondac 50(7): 26–27.
8. Erlandsen, S. L., and E. A. Meyer, editors. 1984. Giardia and giardiasis. New York: Plenum.
9. Bush, A. O., and W. M. Samuel. 1981. A review of helminth communities in beaver (*Castor* spp.) with a survey of *Castor canadensis* in Alberta, Canada. In: J. A. Chapman and D. Pursley, editors. Proceedings of Worldwide Furbearer Conference; 1980 Aug. 3–11; Frostburg, Md. Frostburg, Md.: Worldwide Furbearer Conference, Inc. p 678–689.
10. Joszt, L. 1964. The helminth parasites of the European beaver, *Castor fiber* L., in Poland. Acta Parasitologica Poloniae 12: 85–88.
11. Romashov, V. A. 1969. Helminth fauna of European beaver in its aboriginal colonies of Eurasia. Acta Parasitologica Polonica (Warsaw) 17(7): 55–64.
12. Neumann, V., and R. Piechocki. 1985. Morphologische und histologische Untersuchungen an den Larvenstadien von *Platypsyllus castoris* RITSEMA (Coleoptera, Leptinidae). Entomologische Abhandlungen Staatliches Museum für Tierkunde Dresden 49: 27–34.
13. Fain, A., J. O. Whitaker, and M. A. Smith. 1984. Fur mites of the genus *Schizocarpus* TROUESSART, 1896 (Acari: Chirodiscidae) parasitic on the American beaver *Castor canadensis* in Indiana, USA. Bulletin et Annales de la Societe Royale Belge d'Entomologie 120: 211–239.
14. Whitaker, J. O., Jr., A. Fain, and G. S. Jones. 1989. Ectoparasites from beavers from Massachusetts and Maine. International Journal of Acarology 15: 153–162.
15. Fain, A. and J. O., Whitaker. 1988. Mites of the genus *Schizocarpus* TROUESSART, 1896 (Acari, Chirodiscidae) from Alaska and Indiana, USA. Acarologia 29: 395–409.

chapter 16

Maker of Landscapes: Creating Habitat for Plants, Animals, and People

In places which have been long frequented by beaver undisturbed, their dams, by frequent repairing, become a solid bank, capable of resisting a great force both of water and ice; and as the willow, poplar and birch generally take root and shoot up, they by degree form a kind of regular-planted hedge, which I have seen in some places so tall, that birds have built their nests among the branches.

Samuel Hearne, 1795

It would pay us to keep beaver colonies in the heights. Beaver would help keep America beautiful. A beaver colony in the wilds gives a touch of romance and a rare charm to the outdoors. The works of the beaver have ever intensely interested the human mind. Beaver works may do for children what schools, sermons, companions, and even home sometimes fail to do,—develop the power to think. No boy or girl can become intimately acquainted with the ways of works of these primitive folk without having the eyes of observation opened, and acquiring a permanent interest in the wide world in which we live.

Enos A. Mills, 1913

Beavers profoundly affect their ecosystem by damming up water and removing trees. Needless to say, the stored water and raised water table can be important for many plants and animals, especially during droughts. Furthermore, the water flow pattern is altered, reducing erosion. Larger areas are wetted, and there is more sediment accumulation. As beavers open up forest along streams, they create new landscapes, such as ponds, swamps, and meadows, albeit on a smaller scale than "landscape" as humans see it. The new habitat invites a myriad of plants and animals and leads to complex communities of organisms.

Water and Soil Chemistry

Beavers introduce large amounts of plant material into their impoundments. With the inevitable anoxic decomposition, beaver ponds serve as sinks for organic matter.[1] The rate of methane production is up to 33 times higher in beaver ponds than in free-flowing stream sections.[2] Given the larger area of impoundments and the presence of beaver sites on up to 40% of a stream, beavers may account for up to 92% of the methane production in a river catchment.[2] But as Moore[3] pointed out in 1988, the beaver is not a significant contributor to the 11% increase of methane in the entire troposphere and should not be held co-responsible for the "greenhouse effect" on our climate. Compared to the world's rice paddies, cattle rumens, and fossil fuels, this animal adds little methane. Because of the accumulated organic sediment, the respiration rate is higher in beaver ponds than in free-flowing stream sections. However, the rate of photosynthesis is lower in the impoundments.[4] Beaver ponds also serve as sinks for nitrogen.[5] Nitrogen fixation by sediment microbes is higher in ponds than in unimpounded stream sections.[6,7] The net effect of these chemical processes is a lower pH.

Overall, beavers create "boundary complexity": in the lateral direction, littoral and pelagic areas form a continuum; vertically, the sediment differentiates into upper aerobic and lower anaerobic layers; and longitudinally, the habitat differs between pond or dam on one side, and upstream or downstream reaches on the other.[8]

Habitat for Other Organisms

Plants

Beavers first affect plants by flooding. Small plants will be eliminated, and flooded trees killed within 1 year. Over the longer term, beavers change the vegetation around their pond by cutting down or drowning certain species and creating favorable conditions for others. For instance, in many areas in the northeastern United States beavers have removed over time their preferred tree species and often left behind only evergreens, such as white pine, hemlock, and white spruce. Beavers can influence species composition permanently, even though only small areas along watercourses are affected. At two beaver ponds in Minnesota, selective cutting of trembling aspen reduced the tree density by as much as 43%. Eventually, the avoided species such as alder and conifers dominated.[9] Given beavers' preferences for certain tree species and present forest composition, Barnes and Dibble[10] predicted species composition in a Wisconsin forest after five tree generations: silver maple (*Acer saccharinum*), black birch (*Betula nigra*), and butternut (*Juglans cinerea*) would be completely eliminated, while basswood (*Tilia americana*) and American elm (*Ulmus* sp.) would increase drastically.

In Michigan, beavers had little impact on the number of woody species present near their streams. But plant diversity, taking into account plant dominance, was lower at sites beavers had inhabited for 10–20 years than where beavers had not been allowed to colonize. Especially deciduous trees over 25 cm in diameter were scarce. Beavers' foraging resulted in a higher basal area (the sum of the cross sections of all standing trees) of all woody stems, but the basal area of coniferous trees did not differ from that in sites without beavers. Trees preferred by beavers such as alder (*Alnus rugosa*) and aspen (*Populus tremuloides*) had a larger total basal area where beavers lived. Possibly the animals had stimulated new growth by cutting these trees repeatedly, opening up the canopy and permitting more light through.[11]

In the Netherlands, European beavers appear to change a willow-dominated riparian forest in the direction of less diversity as they selectively cut the rarer nonwillow species to supplement their willow diet.[12]

Similarly, as reintroduced beavers spread to new streams in Bavaria and flooded woodlands, ash thrived around the beaver ponds, while spruce died. Selective cutting changed the mix of trees: near the water, ash, poplar, and elm decreased in relative frequency, and distant from water, willow and poplar trees became less numerous. Dam building raised the water table by 30–50 cm (12–20 inches), which in turn favored certain woody species.[13]

Insects and Other Invertebrates

Aquatic insect populations are greatly affected by beaver activity. Mosquitoes become less numerous in beaver ponds, and the species composition of their populations changes.[14] In general, gravel-loving aquatic insects such as stoneflies will be displaced by silt-dwelling species, such as mayfly and dragonfly larvae.[15] Beavers can even have an impact on terrestrial insects: white pine weevils (*Pissodes strobi*) invaded a white pine (*Pinus strobus*) plantation that had been established in partial shade of bigtooth aspen (*Populus grandidentata*). This happened after beavers cut down the aspen and thus provided ideal conditions for the weevil.[16]

In four boreal headwater lakes in Ontario, Canada, beaver lodges with their decaying wood added greatly to the animal abundance of the otherwise rather barren shoreline of sand and rocks. Only 7 taxa were more abundant in sand and rocks: trichoptera, beetles, oligochaeta, 3 fishes (slimy sculpin, brook stickleback, and fathead minnow), and tadpoles of *Rana clamitans*. By contrast, 14 taxa occurred primarily within 8–10 m from the beaver lodges. Total densities of benthic macroinvertebrates were almost 3 times higher near beaver lodges than at sandy and rocky shorelines. Likewise, the densities of animals in the open water were more than 3 times higher here. On the other hand, only water above rocks harbored crayfish, while beaver sites lacked them.[17]

Beaver impoundments in Swedish streams contained 24 different invertebrate families, compared with only 18 in comparable slow-flowing stream sections.[18]

Fish

Beavers affect fish in both beneficial and detrimental ways. In Missouri, beavers provided a greater volume of water, deep pools during periods of intermittent stream flow, and an increased variety of fish habitat. The standing crop of plankton in beaver ponds was 5 times larger than in the unaltered stream, fish species were more varied and abundant, and their standing crop was larger. In short, the beaver activity increased the stream's carrying capacity for warm-water fishes.[19] Warm-water fish, or mud minnows (Umbridae) increase in numbers in beaver ponds but will die when the water turns toxic. The deep area just above the dam is good habitat for northern pike (*Esox lucius*) and smallmouth bass (*Micropterus domomieui).*[15] Beaver dams may constitute temporary barriers to spawning runs of trout, northern pike, walleyed pike, white suckers (Plate 35), and others. Whether migrating fish manage to traverse the dams depends on the discharge from the pond. In spring, during heavy water flow, fish may surmount the dam more easily.[8] Water flowing around the dam facilitates upstream migration. On the other hand, beaver ponds often provide the last remaining water during severe droughts. Trout may find refuge there when a stream dries up.[15] After droughts or winters, fish recolonize small streams from downstream where the environment is more stable and more fish species occur.

The steeper the stream gradient, the more the fish benefit from the beaver's dam building. Frigid mountain water is spread and warmed to the trout's optimum, while impounded low-gradient streams become too warm and eventually even toxic for trout. Thus, in the American Midwest, for instance, beavers tend to be more detrimental than beneficial to trout.[8]

In the Californian Sierras, beaver dams led to an accumulation of silt covering the gravel, which the native golden trout (*Salmo irideus*) needs for spawning. Brown and rainbow trout displaced this species. The brown trout is larger and cannibalistic, and the rainbow trout hybridizes with golden trout.

Coho salmon use beaver ponds at coastal streams in Oregon, especially during low-water periods in late summer.[20] During summer, the density of juvenile coho salmon (1.43/m^2) in the Fish Creek Basin in the Cascade Mountains of northwest Oregon was 4 times higher than the density in side channels and 48 times higher than that in riffles. Although a beaver pond constituted only 2.5% of the habitat at Fish Creek, it produced 50.4% of the coho salmon smolts in 1986, more than in 1985.[21] In winter, beaver ponds along with alcoves supported most juvenile coho salmon (about 1 fish/m^2) when compared with other stream parts such as backwater pools, trench pools, glides, riffles, and rapids. Alcoves and beaver ponds represented only 9% of the habitat but accounted for 66% of the coho salmon found.[22] Likewise, a series of deserted beaver dams, draining 14 ha, proved to be important winter habitat for coho salmon at a stream on Vancouver Island. Juvenile coho survived better there over the winter (61%–74% survived) than in the entire stream system (35%).[23]

In Swedish streams 2.0–6.5 m wide, minnows spawned in beaver ponds, while brown trout grew larger in beaver ponds but occurred there in lower numbers than in riffles. Overall, beaver ponds permit fish to avoid severe winter conditions and periods of low flow. They are thought to enhance fish species diversity.[24]

As beaver sites transform from active to collapsed ponds, they benefit fish in different ways. "Closed environments" such as active upland beaver ponds tend to have the largest number of fish but few species, while collapsed beaver ponds and streams ("open habitat") harbor the most species of fish. Such open habitat also suffers less oxygen stress during both summer and winter.[25]

Beyond ponds that beavers have sculpted out of streams, beaver lodges add new habitat for fish even in lakes. In lakes in Ontario, 26–57 ha large, northern red-belly and finescale dace (genus *Phoxinus*) occurred in higher densities at beaver lodges than along the rest of the rocky and sandy shoreline.[17]

Amphibians

Beaver ponds harbor salamanders, frogs, and toads. All reproduce there. The spring chorus of tree frogs, the trill of the American toad, and the green frog's banjo-string calls are hallmarks of a thriving beaver site.

Red-spotted newts (*Notophthalmus viridescens*) colonize new ponds rapidly and may actually depend on beaver ponds for their survival. They respond to pond habitats that rapidly shift in space and time. The life history of the red-spotted newt appears particularly adapted to small and temporally unstable bodies of water such as beaver ponds. During its red eft phase, the newt quickly immigrates into newly formed beaver ponds. Beaver ponds last about 10–25 years, but newt populations need 50 years or more to reach saturation. If this is true, newt populations never reach the carrying capacity of a beaver pond.[26]

The wood frog (*Rana sylvatica*), North America's most northerly species of all amphibians and reptiles, breeds and produces in beaver ponds. It uses small, marginal ponds with little inflow and outflow, located in a nonoptimal habitat. Only habitats saturated with beaver populations will have such ponds. Well-managed beaver populations, therefore, will have fewer ponds suited for wood frogs. The metamorphosed froglets emerging from beaver ponds tend to be larger than those from vernal (ephemeral) pools.[27]

Wood frogs taken from beaver ponds and from forested wetlands differed in their development in an outdoor laboratory. First, in shaded containers, embryos from forested wetlands hatched sooner than those from beaver wetlands. In full sunlight, they did not differ. Second, larvae from beaver ponds tolerated higher temperatures than those from forested wetlands. The first fact is interpreted as a need to develop faster in cooler, forested wetlands, before the ephemeral pond dries out. The tolerance of higher temperatures may reflect the sometimes extreme warming of the unshaded beaver ponds—up to 40°C.[28]

In a more southern region (South Carolina), amphibian populations did not differ significantly between streams and new and old beaver ponds. Ten species of salamanders and 10 species of frogs and toads occurred in similar numbers of individuals and species diversity in all three habitats.[29]

Reptiles

Among the reptiles, particularly turtles are attracted to beaver ponds. Painted turtles bask on logs in the water. Snapping turtles hide in the mud and lay their eggs in sandy banks of the ponds. Beaver ponds may also be important for rare and endangered species of reptiles. In South Carolina, more turtles, lizards, and snakes were trapped at beaver ponds that were at least 10 years old than at ponds 5 years old or newer. Streams without beaver structures had even fewer reptiles.[29]

Birds

Water impounded by beavers provides new habitat for waterfowl and many other birds. The long list of birds we observed feeding or breeding at beaver ponds in New York includes American and hooded mergansers, Canada geese (Plate 36), mallards, pintails, buffleheads, wood ducks, horned and pie-billed grebes, great blue and little green herons, kingfishers, woodpeckers, chickadees, tree swallows, eastern bluebirds, red-winged blackbirds, and numerous species of flycatchers and warblers. In the U.S. state of Georgia, dead snags that beavers had created by girdling and flooding attracted more than twice as many woodpeckers of seven species than did a tree stand without beavers.[30] A survey of birds at beaver ponds in eight counties in New York State demonstrated that active beaver sites support more species of birds than do vacant or potential sites.[31] The great blue heron exists in the Adirondack Mountains mostly because of dead standing trees that have been killed by beavers (P. Houlihan, personal communication, 1997). Among the birds, this heron has perhaps benefited most from the proliferation of beavers in the Adirondacks. In Maine, all ponds selected by females of the American black duck (*Anas rubripes*) were within active beaver colonies. Flooded alder-willow thickets, herbaceous vegetation, and large water surfaces attracted the hens. Thus, the still-expanding beaver population benefits black ducks by creating new brood-rearing habitat.[32] Untrapped beaver sites in south central Maine supported more Canada geese, hooded mergansers, and mallards than did trapped areas.

In Wisconsin, beaver ponds attracted mallards, black ducks, blue-winged teal, ring-necked ducks, and hooded mergansers.[15] Shorebirds, swallows, flycatchers, hawks, warblers, sparrows, kingfishers, osprey, and bald eagles appeared at Wisconsin beaver ponds.[15] In the Rocky Mountains, birds such as green-winged teal, mallards, red-winged blackbirds, Brewer's blackbird, common snipe, and spotted sandpipers rely on beaver ponds.[33]

In highly developed countries with very few remaining wetlands, the new ponds and meadows created by reintroduced beavers play an even greater role as habitat for rare animals. For example, in the former East Germany, cranes were attracted to water impounded by reintroduced beavers.

Experimental flooding of a creek in Finland resulted in large populations of water fleas (Cladocera) during the first year of inundation, and midges (Chironomidae) in the second year. These invertebrates served as food for young and adult teals, goldeneyes, and mallards.[34]

Mammals

Mammals found at beaver ponds include otters who use beaver lodges and visit the open water near dams in the winter. Knudsen[15] found that mink use beaver ponds more than the stream sections above and below (Plate 37). The same was true for raccoons, which find many aquatic insects, such as those in the genera *Benacus, Lethocerus, Dytiscus, Hydrous,* and *Hydrophilus.* The ponds also abound with crayfish, small snakes, and minnows, all taken by the raccoon. We found that most beaver ponds are visited by white-tailed deer. Likewise, Knudsen[15] found that 86% of his ponds had deer signs. Black bears wallow and graze in the pond areas. Muskrats build their houses in beaver ponds, particularly in winter. Knudsen[15] found 3 houses/pond, on the average, or 1 muskrat house/acre of pond. The pioneer moose that since 1986 ventured into northern New York State sought out and stayed at beaver meadows (Plate 38). These areas are so important to moose that they have been surveyed statewide to predict how much suitable habitat is available and what the carrying capacity will be.

Humans

When people first encroached on beaver habitat, the beaver meadows were welcome sites to start agriculture. Flat, rich in organic matter, and largely devoid of trees, the bluejoint grass and sedges offered pasture for livestock. The grass grew "as high as man's shoulder."[35] Chroniclers reported harvests of "four tons of hay cut on an acre."[36] After domestic stock had killed the brush, ditches were dug to drain the water, and, over time, plowing converted the beaver meadows into fields. Today, landscapes changed by beavers are either welcome or undesirable. On the whole, however, the added wildlife and increased recreational opportunities such as fishing, hunting, and bird and beaver watching (Plate 39) benefit local residents as well as vacationing visitors.

Succession

As the vegetation changes at beaver sites, different wildlife species will find favorable conditions. Knudsen[15] described these successional phases for Wisconsin. Before impoundment, the floodplains with alder and willow provide ruffed grouse (*Bonasa umbellus*) with "runs" for brood cover, and woodcocks (*Philohela minor*) nest in swampy areas. The stream itself is free of silt and good to excellent for brook trout (*Salvelinus fontinalis*). Shade from overhanging branches of alder, willow, spruce, fir, maple, elm, ash, and others keeps the water cool. After impoundment, muskrats and ducks move in. Older beaver sites, when drained, constitute often the only extensive clearings in vast areas of forest. Such "beaver meadows" provide grazing for deer and moose. Beaver meadows are of low pH, so anaerobic microorganisms produce toxic "swamp gases" such as hydrogen sulfide. This gas affects iron compounds, which in turn fix or "tie up" phosphorus in insoluble form. The direct effects of hydrogen sulfide and the soluble iron compounds damage roots and their mycorrhizal fungi,[37] so that trees cannot grow. But rain and melt water will leach out toxic materials. Woody plants will slowly invade the opening from its edges 5–10 years after the pond has been drained. The invading shrubs shade the watercourse anew, and the lower temperature improves conditions for trout. However, in some areas no trees will grow, and the wetland will persist indefinitely.

In summary then, beaver impoundments adversely affect trout and drown timber locally, while wildlife in general increases in abundance, and recreation for people is enhanced. The effects will vary with local conditions, and while some species will be hindered, others will be helped. For example, in shallow trout streams beavers do more harm to trout habitat than good but create habitat for such animals as ducks and muskrats.

REFERENCES

1. Hodkinson, I. D. 1975. Energy flow and organic matter decomposition in an abandoned beaver pond ecosystem. Oecologia 21: 131–139.
2. Ford, T. E., and R. J. Naiman. 1988. Alteration of carbon cycling by beaver: methane evasion rates from boreal forest streams and rivers. Canadian Journal of Zoology 66: 529–533.
3. Moore, P. D. 1988. The dam busters. Nature 334: 295.
4. Naiman, R. J., J. M. Melillo, and J. E. Hobbie. 1986. Ecosystem alteration of boreal forest streams by beaver (*Castor canadensis*). Ecology 67: 1254–1269.
5. Naiman, R. J., and M. Melillo, 1984. Nitrogen budget of a subarctic stream altered by beaver (*Castor canadensis*). Oecologia 62: 150–155.
6. Naiman, R. J. 1984. The influence of beaver (*Castor*) on the dynamics of lotic ecosystems [abstract]. Presented at 32nd annual meeting of North American Benthological Society, Raleigh, N.C.
7. Francis, M. M., R. J. Naiman, and J. M. Melillo. 1985. Nitrogen fixation in subarctic streams influenced by beaver (*Castor canadensis*). Hydrobiologia 121: 193–202.

8. Schlosser, I. J. 1995. Dispersal, boundary processes, and trophic level interactions in streams adjacent to beaver ponds. Ecology 76: 908–925.
9. Johnston, C. A., and R. J. Naiman. 1990. Browse selection by beaver: effects on riparian forest composition. Canadian Journal of Forestry 20: 1036–1043.
10. Barnes, W. J., and E. Dibble. 1988. The effects of beaver in riverbank forest succession. Canadian Journal of Botany 66: 40–44.
11. Ritchie, M. E. 1983. The impact of selective foraging by beaver on forest community structure. Acta Zoologica Fennica 174: 310.
12. Nolet, B. A., A. Hoekstra, and M. M. Ottenheim. 1994. Selective foraging on woody species by the beaver, *Castor fiber*, and its impact on a riparian willow forest. Biological Conservation 70: 117–128.
13. Zahner, V. 1996. Der Einfluss des Bibers (*Castor fiber*) auf gewässernahe Wälder [doctoral dissertation]. Munich: Munich University.
14. Butts, W. L. 1992. Changes in local mosquito fauna following beaver (*Castor canadensis*) activity—an update. Journal of American Mosquito Control Association 8: 331–332.
15. Knudsen, G. J. 1962. Relationship of beaver to forests, trout and wildlife in Wisconsin. Technical Bulletin 25. Madison: Wisconsin Conservation Department.
16. McNeel, W. 1964. Beaver cuttings in aspen indirectly detrimental to white pine. Journal of Wildlife Management 28: 861–863.
17. France, R. L. 1997. The importance of beaver lodges in structuring littoral communities in boreal headwater lakes. Canadian Journal of Zoology 75: 1009–1013.
18. Sjöberg, G. 1999. Ecosystem engineering in forest streams—invertebrate fauna in beaver ponds [abstract]. European-American Mammal Congress; 1998 July 19–24; Santiago de Compostela, Spain. SY-15, 253. Santiago de Compostela: Universidad de Spain. p 158.
19. Hanson, W. D., and R. S. Campbell. 1963. The effect of pool size and beaver activity on distribution and abundance of warm-water fishes in a North Missouri stream. American Midland Naturalist 169: 136–149.
20. Leidholt-Bruner, K., D. E. Hibbs, and W. C. McComb. 1992. Beaver dam locations and their effects on distribution and abundance of coho salmon fry in two coastal Oregon streams. Northwest Science 66: 218–223.
21. Everest, F. H., G. H. Reeves, J. R. Sedell, D. B. Hohler, and T. Cain. 1987. The effects of habitat enhancement on steelhead trout and coho salmon production, habitat utilization, and habitat availability in Fish Creek, Oregon, 1983–86. Portland, Ore.: U.S. Department of Energy, Bonneville Power Administration.
22. Nickelson, R. R., J. D. Rodgers, S. L. Johnson, and M. F. Solazzi. 1992. Seasonal changes in habitat use by juvenile coho salmon (*Oncorhynchus kisutch*) in Oregon coastal streams. Canadian Journal of Fisheries Aquatic Sciences 49: 783–789.
23. Bustard, D. R., and D. W. Narver. 1975. Aspects of juvenile coho salmon (*Oncorhynchus kisutch*) and steelhead trout (*Salmo gairdneri*). Journal of the Fisheries Board of Canada 32: 667–680.
24. Hägglund, Å., and G. Sjöberg. 1999. Effects of beaver dams on the fish fauna of forest streams. Forest Ecology and Management 115: 259–266.

25. Schlosser, I. J., and L. W. Kallemeyn. 2000. Spatial variation in fish assemblages across a beaver-influenced successional landscape. Ecology 81: 1371–1382.
26. Gill, D. E. 1978. The metapopulation ecology of the red-spotted newt, *Notophthalmus viridescens* (Rafinesque). Ecological Monographs 48: 145–166.
27. Quail, R. A. C. 2001. The importance of beaver ponds to vernal pool breeding amphibians [M.S. thesis]. Syracuse: State University of New York College of Environmental Science and Forestry.
28. Skelly, D. K., and L. K. Freidenburg. 2000. Effects of beaver on the thermal biology of an amphibian. Ecology Letters 3: 483–486.
29. Russell, K. R., C. E. Moorman, J. K. Edwards, B. S. Metts, and D. C. Guynn Jr. 1999. Amphibian and reptile communities associated with beaver (*Castor canadensis*) ponds and unimpounded streams in the Piedmont of South Carolina. Journal of Freshwater Ecology 14: 149–158.
30. Lochmiller, R. L. 1979. Use of beaver ponds by southeastern woodpeckers in winter. Journal of Wildlife Management 43: 263–266.
31. Grover, A. M., and G. A. Baldassarre. 1995. Bird species richness within beaver ponds in south-central New York. Wetlands 15: 108–118.
32. McCall, T. C., T. P. Hodgman, and D. R. Diefenbach. 1996. Beaver populations and their relation to wetland habitat and breeding waterfowl in Maine. Wetlands 16: 163–172.
33. Brown, D. J., W. A. Hubert, and S. H. Anderson. 1996. Beaver ponds create wetland habitat for birds in mountains of southeastern Wyoming. Wetlands 16: 127–133.
34. Nummi, P. 1992. The importance of beaver ponds to water fowl broods: an experiment and natural tests. Annales Zoologici Fennici 29: 47–55.
35. Cronon, W. 1983. Changes in the land: Indians, colonists, and the ecology of New England. New York: Hill and Wang.
36. Whitney, G. G. 1994. From coastal wilderness to fruited plain: a history of environmental change in temperate North America, 1500 to the present. Cambridge, UK: Cambridge University Press.
37. Wilde, S. A., C. T. Youngberg, and J. H. Hovind. 1950. Changes in composition of ground water, soil fertility, and forest growth produced by the construction and removal of beaver dams. Journal of Wildlife Management 14: 123–128.

Part V

Beaver and People: Conservation, Use, and Management

chapter 17

"Here before Christ": Fur Trade, the "Beaver Republic" (Hudson's Bay Company), and Fur Trapping Today

> The beaver makes everything perfectly well, it makes us kettles, swords, knives, bread; in short, it makes everything . . . without the trouble of cultivating the ground. The English have no sense; they give us twenty knives . . . for one beaver skin.
>
> *A Montagnais Indian, 1634*

No other wild animal has shaped history as much as the beaver. It has lured fur trappers and traders more and more deeply into the northern wilderness for two centuries, from the mid-1600s to the late 1850s. The fur trade painted the map of North America's interior and paved the way for European settlement, the founding of empires, and the destruction of indigenous cultures. After trapping out the tributaries of the St. Lawrence River, Indian and half-caste fur trappers and traders pushed westward and northward, eventually exploiting the Rocky Mountains and reaching the Pacific Ocean. The beaver trade's best-known and most important formal business organization was the Hudson's Bay Company.

The Beginnings: 15th and 16th Centuries

John Cabot, a Venetian immigrant to England, made landfall at the east coast of either Newfoundland or Cape Breton Island in 1497. The following year Cabot led settlers from Bristol to the New World in five ships and planned to continue to sail to Japan (Chipangu). All ships but his sank.

From 1534 to 1542 Jacques Cartier from St.-Malo explored the area of the mouth of the St. Lawrence River. In Chaleur Bay between Quebec and New Brunswick, Micmac Indians offered Cartier furs for beads and knives and traded away even the robes off their bodies. Cartier searched for a route to China. Rapids near the Indian village Hochelaga, which means "Beaver Meadows," blocked his way up the St. Lawrence River. Today Hochelaga is Montreal, in the past called the "queen of the beaver trade."

Sailors, sealers, and early explorers working off Newfoundland started the fur trade with casual exchanges of manufactured goods from Europe for furs from the Indians in the late 1500s. Norman codfishermen from St.-Malo, Rouen, and Dieppe on France's north coast, in order to prevent Indians from pilfering their cod-drying platforms on rocks, started to trade with them. They bartered beaver

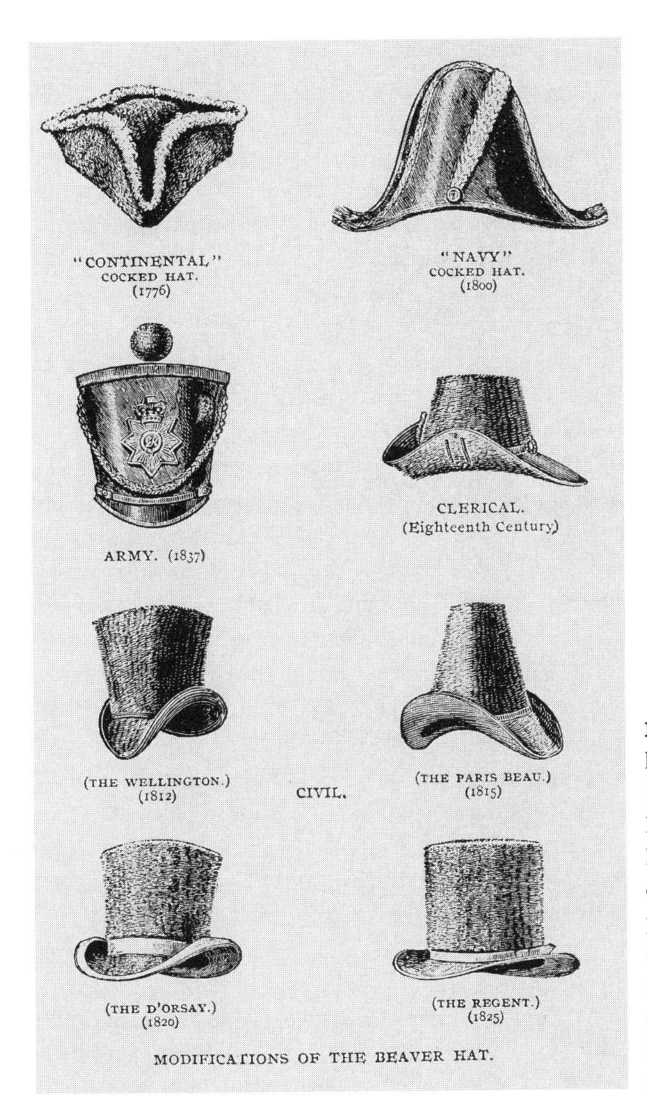

Figure 17.1 | Beaver hats of the 18th and 19th centuries. (From H. V. Radford. 1904–06. History of the Adirondack beaver. Annual reports of the Forest, Fish and Game Commission of the state of New York. Albany, N.Y.: the Forest, Fish and Game Commission. p 389–418.)

pelts for fishermen's knives. The Europeans at the time used the fur for many purposes, especially beaver felt hats (Fig. 17.1) made in Flanders. In the 16th century fur had become scarce in Europe, after fur from marten had been "in" during the previous century. Before that, in the 13th and 14th centuries, the English had been using squirrel for fur.[1]

The Basques also brought beaver furs, along with whale oil, to their homeport of St. Jean de Luz on the French side of the Bay of Biscay. In 1578, as many as 350 vessels fished the Grand Banks off Newfoundland. They were English, French, Spanish, and Portuguese.

To the explorers seeking a northwest passage to Cathay (China) and the Spice Islands of South East Asia, the Arctic seemed desolate. They could not imagine the wealth of furs to be obtained there. Starting with John Cabot's voyage to Newfoundland in 1497, the search was continued by Verrazano and Cartier in the 16th century. During the reign of Queen Elizabeth I (1558–1603), several expeditions tried to find a northwest passage. In 1577 Martin Frobisher explored for the Cathay Company, and John Davis followed in 1585.

17th Century: Beaver Wars

In 1603 Samuel de Champlain, known as the "Father of New France," became the "Forts Governor of Canada." In 1608 he established Quebec as a fur trade base, and in 1643 Montreal followed. Champlain and his associates bought 22,000 skins in 1626, for $2 each. Their fur company paid a dividend of 40%. In 1627, Cardinal Richelieu of France established the Company of New France, also called the "Company of 100 Associates" (in French: Compagnie des Cent-Associés), with himself as leader. The company was granted the entire St. Lawrence Valley, which included a fur trade monopoly for New France and Acadia, broad governmental powers, license to trade guns to Christian Indians, and an understanding to bring in large numbers of colonists. However, their first expedition of 18 ships in 1628 was captured by 3 Scottish vessels and taken to England. In the Peace Treaty of St. Germain-en-Laye (1632), Canada was returned to France.

Henry Hudson explored the area for the Dutch East India Company in 1607, and after that, William Baffin and Robert Bylot, in 1615. Hudson sailed 150 miles up the Manna-hata River (later to bear his name) in his little Dutch ship *Half Moon* (the first European ship on the river) as far as today's Albany, named Fort Orange at the time. The town growing up around it was appropriately called Beverwyck. A member of Hudson's crew noted in his diary that "the people of the country came flocking aboard and brought us grapes and Pompions [pumpkins], which we bought for trifles. And many brought us beaver skins, and otter skins, which we bought for beads, knives, and hatchets." These "people of the country" could have been none other than the Haudenosaunee, better known as the Iroquois Confederacy, and specifically its eastern Mohawk nation. The locals knew the craving of Europeans for fur, since the Algonquins along the St. Lawrence and Saguenay Rivers to the north had been trading for furs with French ships since the 1540s. From here on, the competition between European powers would aggravate the preexisting frictions between the Iroquois and Algonquin, as the former allied themselves with the British and Dutch and the latter with the French. In 1608 Champlain and 60 Algonquins (Hurons and Montagnais) canoed south to the lake that now bears his name. A bloody encounter on the lake brought upon the French and their Algonquin allies the enmity of the Iroquois Federacy that would change history.[2]

Hudson's 1607 expedition led to the acquisition in 1626 of Manhattan Island from the Canarsee Indians in exchange for cloth and trinkets valued at $24. On his next trip, searching for a route to the Orient, he reached the south end of Hudson's Bay in 1610. In the winter of 1610/11, an Indian offered two deer and a pair of beaver pelts. Hudson traded a knife, a looking glass, and a few buttons for the skins. Thus started the fur trade at the bay.

During the first quarter of the 17th century, fishermen traded for fur along the New England coast; the French, along the Maine coast; the Dutch, in the south along Long Island Sound and Narragansett Bay; and the English, between these areas.

For the next 100 years the Iroquois fought to wrest the fur trade from the French, who had been shipping the precious cargo down the St. Lawrence River. In 1642 the Iroquois attacked and almost destroyed the Huron nation, once the wealthiest and the most numerous Indians in the Great Lakes region. In this great competition, the Iroquois raided along the St. Lawrence and destroyed the Petun, Neutral, Tobacco, and Erie nations. As these tribes vanished, the French started to trade with the next tribes west. At that time, only about 300 Europeans lived in all of New France, which encompassed Canada and most of the Mississippi Valley (similarly, in 1643 Manhattan had only about 500 residents, speaking 18 languages).[2]

Harvesting beavers and other game required large hunting parties. Before metal traps became available in the late 1700s, the Iroquois broke down dams and speared beavers as they struggled to flee the drained pond. In winter, they cut holes in the ice, destroyed the lodge, and intercepted the beavers as they surfaced for air.[2]

The fur trade motivated the first settlements in North America. After exploratory fur-trading trips on the Maine coast in the first decade of the 17th century, the Pilgrim fathers founded Plymouth Colony in Massachusetts Bay in 1620. It served as a fur-trading post from the outset. As soon as 1621 the ship *Fortune* took beaver skins to Europe, constituting 69% of the cargo's value. The beaver now became the basis of exchange with Europe. The Pilgrims never were successful in the fishing business, so they concentrated on the beaver trade. By 1625 they had turned from the coast to the interior via the Kennebec River. From 1628, and for the next 20–30 years, the Pilgrims traded with Indians in Maine for beavers. They succeeded because they now paid in wampum—shell pieces—they had acquired from the Dutch the year before. By 1630 the English forced the French back to the St. Lawrence River and the Dutch back to the Hudson. While the beaver trade expanded to the interior of Maine, it declined at the coast around 1640. But by that time the Plymouth Colony had profited enough from 20 to 30 years of fur trade to buy essentials in England, pay off their debts, and invest profits in the next enterprise: raising cattle.[3]

From 1625 on, the fur and rum trade with Indians and missionary activity by French Jesuits flourished. After 1674, promotion of the first French colony in the New World was intensified, and by 1690 there were more than 10,000 settlers in Canada.

Members of the Massachusetts Bay Colony also ventured west to exploit new beaver territory. William Pynchon, one of the original patentees, founded Springfield as a fur-trading post in 1635. He dominated the fur trade of the Connecticut River Valley for 16 years (1636–1652) and became one of the richest men in New England. His son John Pynchon built up Springfield while his father returned to England. From 1652 to 1658 he shipped more than 9000 beaver skins to England, and around 1657 over 2000 beaver pelts per year. Nearby Windsor and Hartford were other fur-trading posts. Today's Concord, Massachusetts, started as a fur-trading post, founded by Simon Willard in 1635. As the fur traders pushed west toward the upper Hudson, the rivalry over their trade led to confrontation at the Delaware and Hudson Rivers and the expulsion of the Dutch from the eastern region. The French ventured up the St. Lawrence to the Great Lakes, and the Dutch followed the Hudson River for new beaver grounds. But the English at the Massachusetts and Connecticut coasts had no direct river access over the Appalachians into the interior.

The two French explorers Pierre-Esprit Radisson and Médard Chouart, Sieur des Groseilliers explored between 1654 and 1661 southwest of Lake Superior and discovered the Mississippi. Around 1666 Radisson and Groseilliers ("gooseberry") planned to build trading posts for Hudson's Bay and for beaver pelt shipments to France but were rebuffed by France. They offered their plan to the English, suggesting that they could bypass French posts on the St. Lawrence River by trading furs through Hudson's Bay.[4,5] This idea excited Prince Rupert, Count Palatine of the Rhine and Duke of Bavaria, and nephew of King Charles I of England (king from 1625 to 1642). After Groseilliers explored Hudson's Bay in 1668/69, Charles II of England granted Prince Rupert and others a charter as "true and absolute Lordes and Proprietors" of all seas and lands that drain into the Hudson's Bay ("Rupert's Land") on May 2, 1670. Rupert's Land ran from the northern tip of Labrador in the east to the Rocky Mountain ranges west of today's Lethbridge, Calgary, and Edmonton, and from Baffin Island in the north to Lake Superior in the south (Fig. 17.2). Officially, the enterprise was known as "The Governor and Company of Adventurers of England Tradeing into Hudson's Bay."[5] The Hudson's Bay Company (HBC) was born. The sweeping rights to the entire watershed draining into Hudson's Bay were granted under the condition to keep looking vigorously for the northwest passage. Significantly, a trade monopoly for all areas west of Hudson's Bay was granted, so that the putative northwest passage and the trade route all the way to China would have been included.

For many years the search for a northwest passage and commercial interest in

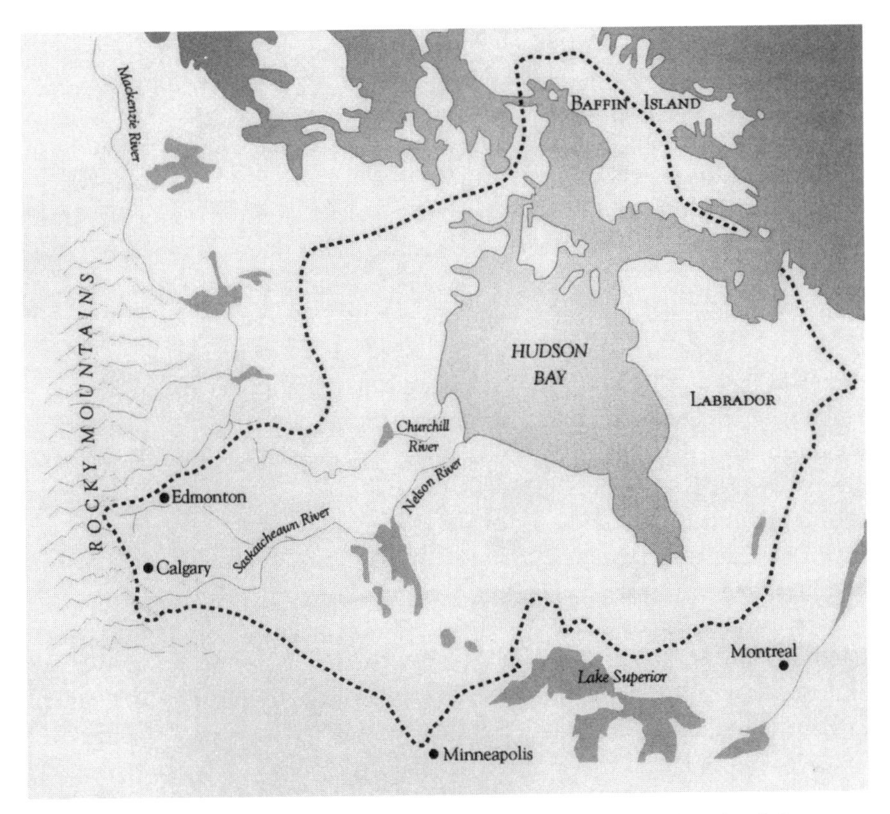

Figure 17.2 | Prince Rupert's Land: the area granted to the original Hudson's Bay Company.

furs were twin motivations for arctic exploration. However, the fur trade had proved to be so profitable that it soon eclipsed the geographical exploration and became the HBC's main pursuit. The company had its headquarters in London, and a governor, deputy governor, and committee represented the shareholders. A North American governor linked them with the field. He in turn was responsible for several departments. In each department, chief factors, or agents, were running larger posts, and chief traders and clerks smaller ones. Express canoes that took only passengers and mail connected all of these outposts.[6] In his book *Company of Adventurers*, Peter C. Newman[7] called it an "independent beaver republic" instead of a business firm. The beaver adorned its coat of arms (Plate 40). The early HBC was a political empire that had its own navies, armies, medals, and currency and even used a calendar that started with the date of its own founding. It was so powerful that nobody could imagine a world without it. Its initials came to mean to many "Here before Christ."

Beaver felt hats were in great demand during the booming times of the 17th century in England, fueling the fur trade. At that time, Christopher Wren was rebuilding London after the fire of 1666. Arts and sciences thrived. In England, Milton was dictating *Paradise Lost* (1667), and Newton discovered the laws of gravity. On the continent, it was the time of Vermeer, Rembrandt, and Spinoza; Velázquez and Murillo; Molière; and Heinrich Schütz and Leibniz. In 1638 King Charles I of England had directed that "nothing but beaver stuff or beaver wool shall be used in the making of hats."[2] Beaver hats were made from the felt of the beaver's thick, soft, and barbed underhair that was shaved from the skin. What follows is a description of the steps in making a hat from beaver felt. These beaver hats originally served as rain gear but soon became an item of fashion that distinguished social classes.

How to Make a Beaver Hat

Remove the long guard hairs from the pelt.
Shave off the short hair. It looks like down.
Nine ounces of it is needed for a hat.
Fluff it up with a "vibrating bow": the hair is being "bowed."
Push down with a "hatter's basket" to even it out.
Roll into a sheet of wet linen. Now the barbs of the hair hook together.
Now it looks like felt or like a giant patch of lint from a washing machine.
Roll it up.
Two rolled-up squares ("batts") are needed for one hat.
Two layers of batts are folded over a cone-shaped paper form.
Rolling the felt into two directions locks the fibers together.
Hot water bath (262°F) with sulfuric acid shrinks and strengthens felt.
(Formerly mercury compounds were used for this. These compounds caused poisoning, the "madhatter's disease.")

Place onto a wooden form and crowning tool.
Shape into a hat, tied with string around the base of the cone.
Trim the edge.
Orange shellac flattens and stiffens the brim.
Felt surface is smoothed with pumice, brush, and knife.
Tallow makes the hat water-resistant and lustrous.
Hat brim is ironed with a tiny flatiron.
Hat can be died any color.
Decorate with ribbon and turkey feather.

Wish for a beaver hat? They are still being custom-made, although only in modern versions, with tools as old as 300 years. See, for example www.davebrownhats.com.

On the way from the Native trapper to the fur merchant in Europe, the price for a beaver pelt was marked up by almost 1000%. Since the coarse guard hair was difficult to remove from the fur, Native trappers often fashioned several pelts into a rather basic coat and wore it for several seasons until the guard hair fell out. Now the pelt was a "coat beaver" or, in French, *castor gras* and more valuable than a regular "parchment" or *castor sec.*[2]

With the founding of the HBC in 1670, rivalry between the British and French fur empires developed. The British sought to squeeze the French in their St. Lawrence domain. They advanced north from the Hudson River area and south from Hudson's Bay. They also had founded Charleston, South Carolina, in 1670, adding to the pressure from the south. In New York, the demand for furs exceeded the harvest by traditional methods, and by 1670 most furbearers were gone. The Iroquois expanded their realm, soon reaching from the Carolinas to the Mississippi Valley and Hudson's Bay. Around 1700 they regularly traveled 1000 miles in the fall to procure furs they would trade in the spring at Albany. Albany held the fur-trading monopoly in New York until 1725, when a trading post opened at Oswego on Lake Ontario.[2]

Furs were important trade goods until about 1675. From then on trade with the West Indies offered new opportunities for shipping fish and lumber. Still, in the late 17th century tension over the fur trade between the French and British intensified. Charles II died in 1685, and the French did not take his successor, his brother James II, seriously. Led by Chevalier Pierre de Troyes, the French raided fur posts at James Bay in 1686: Troyes and Pierre Le Moyne (Seignorial name: Sieur d'Iberville) traveled from Montreal via the Ottawa and Moose Rivers to Moose Factory and seized three posts and their furs. (*Factory* referred to the "factor," the company's agent who lived there). More fur raids occurred in 1688 and 1689. King James was driven from England in 1688, and William of Orange became king. On May 17, 1689, war was declared known as "King William's War." In 1697 while England and France were officially at war, five French ships sailed from

La Rochelle in Normandy on another fur raid. After a sea battle with three English ships in Hudson Strait, the French took over York Factory, HBC's main outpost on Hudson's Bay. York Factory, located on the Hayes River, placed the HBC 1500 miles closer to the source of furs in the West than the French, with headquarters back in Montreal, known as "Montreal Traders." These "Beaver Wars" ended in 1713 with the Peace Treaty of Utrecht when France surrendered all rights to Hudson's Bay and ceded Newfoundland and Acadia (Nova Scotia) to England. The forts were handed back to England in 1714. Yet, in 1782, French marines from the West Indies sacked York Factory.

After 1685 more furs were smuggled from Canada (the French sphere) to New York than were brought in legally from the West. This was because fur traders preferred most manufactured goods from the British. Of French wares, only gunpowder was deemed desirable. Converted Iroquois, living near Montreal, carried the pelts for the Canadians through the mountains of the Adirondacks. They were the Caughnawagas, "Praying Indians."[2] (The Caughnawaga—now spelled Kahnawake—Reserve still exists in Quebec Province, across the St. Lawrence River from Montreal). To evade the French, the New York traders knew their Canadian contacts only by code and never learned with whom they were dealing.[2] When beavers became scarce, the Iroquois earned more as smugglers than as trappers because they knew their woods so well.

Societies changed as Eskimos and Indians grew dependent on fur trapping, the vagaries of demand and prices, and the power of the HBC, and ultimately suffered from the depletion of the furbearer stocks.

The HBC used the beaver skin as a standard for bartering and virtually as currency. (At the time, paper currencies in the North American colonies suffered much hyperinflation.) One adult-size, prime-quality beaver skin ("made-beaver," M-B) was worth 1 brass kettle, or 2 lb of sugar, 1 gallon of brandy, 1 blanket, 20 fishhooks, 2 pairs of goggles, 2 hatchets, 1 ice chisel, 8 knives, 4 spoons, or 2 shirts. Most popular were guns, costing 132 beaver skins each. Other furs or animal products were quoted in made-beaver equivalents. In 1703, 1 otter skin, 1 black bear, 2 foxes, 2 woodchucks, 4 raccoons, 8 pairs of moose hooves, or 5 lb of goose feathers were worth 1 made-beaver; a moose hide was worth 2. In 1733 the HBC post at Albany Fort listed the value of one prime-quality adult beaver skin. In bartering, 1 made-beaver fetched ¾ lb of colored beads, or 1 brass kettle, or 1 ½ lb of gunpowder, or 2 lb of sugar, or 2 lb of Brazil tobacco, or 1 gallon of brandy, or 1 blanket, or 12-dozen buttons. The HBC's "point blankets" had short indigo lines or "points" woven into one edge to indicate the number of beaver pelts each was worth—up to about 7. Woven in England of pure virgin wool, they were introduced in 1779. Today such point blankets can be purchased from the L.L. Bean Company in Maine.

The first vessel ever to sail the Great Lakes was built to serve the fur business as a floating trading post and fort. Assembled in 1679 by René-Robert Cavelier,

Figure 17.3 | Historical marker in New Orleans, Louisiana, commemorating La Salle's claim of the Mississippi watershed for France.

Sieur de la Salle, above the Niagara Falls, the 45-ton schooner was named *Griffon.* She sailed across Lakes Erie, Huron, and Michigan, delivered supplies and building materials for a second ship at Green Bay, and picked up furs. She never arrived "back East" with this first cargo of furs, and the wreck has never been found. La Salle, who had stayed behind at Green Bay, later successfully navigated the Mississippi to its mouth. He named the vast territory he had traversed Louisiana, for King Louis XIV of France, who had granted him permission to build the *Griffon* and explore the western wilderness as well as the Mississippi River. After his explorations, La Salle incorporated the Louisiana territory into the French possessions in the New World in 1682. A marker in New Orleans, Louisiana, commemorates this seminal event (Fig. 17.3).

But in the end, his rebellious crew murdered La Salle. Earlier, the Sulpician Fathers had granted La Salle a site on the St. Lawrence River, 6 miles above Montreal. Convinced he was on his way to the orient, La Salle had named the small settlement La Chine. It is today's Lachine, Quebec.

18th Century: French and Indian War

The early 18th century saw the hostilities between the French and British fur interests continue. By 1700 New France outdid the English in the fur trade. Voyageurs, also known as *coureurs-de-bois,* the "lawless brotherhood of the wilderness, scorning all the attempted restraints of Church and State,"[4] traded with Native Americans in the Rocky Mountains, beyond the influence even of the mighty Iroquois. Under pressure from the British, closing in on them from the south and east, the Iroquois Federation made peace with tribes to the north and west in 1701 at a conference in Montreal. They promised to remain neutral in any future war between England and France and allowed Jesuit missionaries into their country. In 1709 Montreal traders and Mohawk Indians attacked Albany Fort on James Bay. The Treaty of Utrecht in 1713 restored order. (By the way, this same Treaty of

Utrecht also transferred Gibraltar from Spain to Britain.) By 1750 the Hidatsa villages on the upper Missouri (above today's Bismarck, North Dakota) had both French and Hudson's Bay representatives in residence each summer.

During the middle of the 18th century the British increased pressure on the French in North America. The French and Indian War from 1756 to 1763 developed from border conflicts between the British and French in the Ohio Valley. In 1758 the British seized Fort Louisburg east of Cape Breton and Fort Duquesne, today's Pittsburgh. In 1759 the British advanced to the St. Lawrence River and took Quebec, and in 1760, Montreal. Now the Great Lakes were accessible to the British. In the Peace of Paris (1763) France lost Canada, Louisiana, and Cape Breton to Great Britain. In Europe this conflict between England and France is known as the Seven-Years War (1756–1763).

During its first century, the HBC competed with French traders and explorers. In the late 1770s the North West Company, with headquarters in Montreal, was founded, and both companies pushed west. The North West Company extracted beaver furs from the Saskatchewan rivers, and from as far as the Arctic Circle and Oregon. To boost the Far West fur trade, the North West Company established the Athabasca and English River Forts. However, with their headquarters at Montreal, it was at a disadvantage. Roundtrips by canoe could not be completed in one season before the rivers froze. Therefore, they had to store trade goods for 2 years. Such handicaps were the downfall of the North West Company. With their forts at Hudson's Bay, the HBC, by contrast, was 1500 miles closer to the beaver grounds of western Canada such as the Lake Winnipeg streams. But the North West Company had more men. In 1799, they numbered 1276, compared with the HBC's 498 employees in North America, 180 of them on the bay.[5] In 1795, the "XY Company" (after the sign on their furs), with Alexander MacKenzie, split from the Northwest Company, only to return to the fold 10 years later.

Seeking refuge from the French Revolution, a French nobleman bought 200,000 acres in northern New York in 1792 and named the enterprise Castorland Corporation. Its coat of arms showed a beaver below the goddess Ceres, a symbol for earth inhabited or cultivated, tapping a maple tree for sugar sap. Castorland failed in 1814. Some of its land became the summer home of Joseph Bonaparte (1768–1844), brother of Napoleon, during his 17 years in America after he had lost the throne of Spain.[2] Lake Bonaparte (New York) reminds of this episode.

19th Century: Ecological Warfare

The still-young United States more than doubled its area by the Louisiana Purchase in 1803. "Louisiana" was a vast area west of the Mississippi and Missouri Rivers, reaching from Canada to the Gulf of Mexico. President Thomas Jefferson, under pressure from his critics, envisioned Louisiana to be peopled by Indians, especially those in the way of expansion by the Euro-American community of settlers, and fur trappers, and not by colonizing American farmers.

Before the Lewis and Clark Expedition (1804–06), the St. Louis fur trade was limited to the lower Missouri. But even before Lewis and Clark returned in 1806, American fur traders ventured upstream to the Mandan villages, intruding in the trading territory of the North West Company.

Sent out by President Jefferson, the Lewis and Clark Expedition sought a northern tributary to the Missouri River above 49° north (border with Canada) in order to wrest at least some of the northern fur trade from the British.[8,9] The American fur trade route via the Missouri, Mississippi, and Ohio Rivers to the saltwater ports of the Atlantic coast (and on to Europe) would be ice-free and faster than the Canadian canoe route. There the canoes had to travel from Lake of the Woods via Rainy Lake to the Great Lakes and the St. Lawrence River. Told by the Mandan Indians about a river flowing into the Missouri from the north, Lewis and Clark believed the White Earth River might be the desired link to the north. But it turned out to be a minor creek. One member of the expedition, John Colter, left it and went to trap beaver. He discovered Yellowstone, which became a national park in 1872.

Other priorities of the expedition included finding an east-west trade route ("the most direct and practicable water communication across the continent, for the purposes of commerce"). This route would use the Columbia River and facilitate trade with China. The expedition also emphasized exploring the abundance of fur animals, as well as the attitude of the native peoples toward the fur trade in the entire area traversed. All of this served to assert American dominance from coast to coast at the expense of the British and Spanish. The fur trade would pave the way for American settlement of the West.

Lewis and Clark carefully noted the number of beavers they sighted and the number caught by themselves and by the French trappers they encountered, as well as the beaver lodges, dams, beaver-cut tree stumps, and the trees suited for beavers such as cottonwoods and willow. Above the Mandan villages on the Missouri River (just north of today's Bismarck) beavers became plentiful. They described the headwaters of the Missouri where the Great Plains meet the Rocky Mountains as "richer in beaver and otter than any country on earth."

John Jacob Astor, an immigrant from Walldorf in southern Germany, entered the fray. He soon became the richest man in America—some say even the world. After immigrating to New York via England, he held a job cleaning and packing furs for auction in London. Sent on business to England, Astor learned that beaver furs brought 900% profit. He founded the American Fur Company in 1809, the Pacific Fur Company in 1810, and the Southwest Company in 1811. He established the fur base Astoria at the mouth of the Columbia River. From there he ran a trade triangle: furs and ginseng from the Northwest coast to China

(30,573 furs in 1800); tea, spices, and silk from China to New York; and blankets, beads, and trinkets from New York to the Northwest coast. His flagship was, appropriately, the *Beaver.* After getting rich on furs, he made an even greater fortune on foreclosures of New York farms deep in debt. His great-grandson William Waldorf Astor built the Waldorf-Astoria Hotel in New York City in 1897. Today, beavers adorn tiles in the subway station at Astor Place in Manhattan (Plate 41).[10]

Between the time of the Lewis and Clark Expedition and 1840, American fur trappers and traders penetrated the trans-Missouri west to the Pacific Ocean and the Gila River in Mexico to the south. Having merged with its competitor, the North West Company of Montreal in 1821, the HBC gained new territory in the Northwest and named it the Columbia Department. It included much of today's British Columbia, Washington, and Oregon. Overall, the HBC now controlled 3 million square miles, an area of about the size of Australia. It was the largest corporate landowner ever.[5]

Afraid of the United States's control of the Pacific Northwest, the company launched a scorched-earth policy in the early 1820s to exhaustively trap all valuable fur animals, mainly beavers. They believed that the value of the land to immigrants would be greatly depreciated without beavers. Thorough extirpation of beavers was to create a "fur desert" or *cordon sanitaire.* About the colonizing Americans, the HBC governor, Sir George Simpson, bluntly stated, "If the country becomes exhausted in Fur bearing animals they can have no inducement to proceed hither."[6] One example of this ecological warfare was the party led in 1826 by Alexander McLeod that traveled from Fort Vancouver on the Columbia River by boat up the Willamette River, then crossed over the coastal mountains of Oregon. At the Pacific coast, the party split, exploring north and south along the coast. In the north, few beavers were found, while the populations to the south were better. The party returned to Fort Vancouver with 285 beaver pelts, and 36 river otter, and 3 sea otter skins. On information given by the Indians, the party returned to southern coastal Oregon the following fall and winter and bagged 797 skins, of which 663 were beaver.[11] The pursuit of beavers grew more intense with the start of yearlong trapping expeditions. In 1825 Governor Simpson stated about the Snake River territory, "The country is a rich preserve of beaver and which for political reasons we should destroy as early as possible."[6] This intentional extirpation of beaver populations in the Pacific Northwest succeeded in less than 2 decades. Between 1826 and 1834 about 3000 beavers were killed every year. But by 1850, trappers in this region could only find 438 beavers a year.[12] During the same decades, between 1825 and 1849, the sustainable trapping system of the Native families in British Columbia, then "New Caledonia," yielded many more beavers than the "political" trapping farther south. These families owned the beaver ponds.[6]

In the 1823/24 season 20 trappers harvested 5000 beavers in 212 traps in the

Bitterroot Mountains alone. One trapper, James O. Pattie, took 250 beavers in 2 weeks on the San Francisco River in Arizona in 1825.[1]

Hunters and trappers sometimes boasted extraordinary numbers of animals bagged in a lifetime. John Hutchins of Manlius, New York, at the age of 64, declared that he had over the years "caught in traps, or otherwise destroyed . . . 100 moose, 1000 deer, 10 caribou, 100 bears, 50 wolves, 500 foxes, 100 raccoons, 25 wild cats, 100 lynx, 150 otter, 600 beaver, 400 fishers, mink and marten by the thousands, muskrats by the ten thousands."[1] These round numbers suggest less than precise record keeping and a bit of hyperbole. Nevertheless, they show the open-ended exploitation of a resource deemed limitless.

The beaver trade declined when the stocks dwindled rapidly after Sewell Newhouse invented the efficient steel trap in Oneida, New York, in 1823. In Europe, silk velour appeared as a lower-priced substitute for beaver felt, reducing the demand. The silk hat was firmly established by 1845. Still, from 1853 to 1877 the HBC sold 3 million skins. York Factory formed the hub of the enterprise: all goods coming in and all furs going out from the western dominion moved through here.[5]

In the West, the last annual Rendevouz, a gathering of trappers and traders, was held in 1849 at Horse Creek on the Green River in Wyoming. Today many place names remind us of the days of the beaver fur trade. Between Rendevouz's, trappers used to cache furs and supplies at safe and dry places. Cache Valley in Utah, today site of Utah State University, derives its name from having been such an important cache site.

More beaver pelts were collected than were skins from mink, marten, lynx, and fox combined, the next most numerous furbearers. Between 1769 and 1868 the HBC auctioned off in London a total of 4,708,702 beaver pelts, on average about 50,000 pelts per year. In addition, the Canada Company and the North West Company sold similarly large numbers. In 1854 alone, 509,000 pelts were sold in London. The HBC auctioned off 3 million skins between 1853 and 1877. The record year for the HBC was 1875, when 270,903 pelts were traded.

Reflecting the decline of the fur trade, the HBC ceded "Rupert's Land," the entire watershed of Hudson's Bay, in a "deed of surrender" to jurisdiction of the Canadian government in 1870. In a sort of final act, the HBC sold its 178 northern stores and its fur auction houses in Toronto and London in 1987 after 317 years of operation. It now concentrates on its real estate operations and its retail business in the population centers, in stores known as "The Bay."

Thus, the political and commercial development of the American north is based on the beaver as on no other animal. No wonder that it adorns the coats of arms of Canada, New York City, and, of course, the HBC. *The Beaver* is Canada's leading history magazine; the animal adorns the Canadian 5-cent coin, the nickel; and the youngest boy scouts are called Beavers. New York and Oregon chose the beaver as their official state animal. Other fur animals such as marten, fisher, otter, mink, and muskrat paled in their importance compared to beavers.

Beaver Trapping Today

Although beavers are still being harvested for their pelts, the fur prices are so low that the main motivations for trapping today are recreation, removal of "nuisance beavers," and managing beaver numbers to prevent them from becoming a nuisance (Fig. 17.4). In North America, where about 20 million beavers live, states or provinces differ in their stipulations. For instance, in New York beavers are trapped in body-gripping ("Conibear") or leghold traps. The latter may not have teeth in the jaws, and padding is recommended. On land, leghold traps cannot exceed 5 ¾ inches in their spread of the jaws, to avoid damage to nontarget species. Under water, they may open 7 ¼ inches wide. Snares are illegal, except by special permission to trained personnel to deal with nuisance beavers. For humane reasons,

Figure 17.4 | A modern-day beaver trapper.

managers prefer padded traps. In Scandinavia, beavers are shot with rifles, not trapped.[13] The hunting season in Sweden lasts from October until May, but most beavers are shot during the last 2 weeks of the season, because of the weather and other, larger game mammals competing for the hunters' time. The annual harvest is about 6000 beavers in Sweden, 5000 in Norway, and 2000 in Finland.[13]

Advocacy of animal rights in general, and specific antitrapping sentiment prevalent in urban populations, have decreased the demand for fur, depressed prices, and exerted pressure on trappers and wildlife managers to use humane methods, prevent accidents to nontarget species, or avoid killing beavers altogether.

Recovery of a population to the point where harvesting can be allowed, indeed may be advisable for population control, can be rapid. Because beaver stocks had declined since the late 19th century, the Province of Ontario, Canada, closed beaver trapping entirely in 1938, with stiff penalties for violations. By the late 1940s the beaver stocks had recovered. Trapping along registered traplines, with exclusive rights and annual quota, started again in 1947, augmented by restocking. About 120,000 beavers were caught annually in the 1950s; 150,000, in the 1960s; and 209,000, in 1979/80 when 17,779 licenses were sold.[14]

In the United States, during the 1963/64 season, 191,248 beavers were trapped in 37 states. Louisiana reported the least (25) and Minnesota, the most (21,600). In Canada, the Province of Manitoba alone boasted 1 million beavers, almost one beaver for every inhabitant. In response to the antifur movement, the number of beavers trapped in Canada has dropped by half to 300,000/year during the last 15 years.[15] Today the price for a beaver pelt that a trapper receives in the United States ranges from $5 for a southern beaver in Louisiana to $30–$40 in the North, but even in Canada prices can drop to as little as $12. Furs are stretched on hoops for drying (Plate 42) and regularly auctioned. A beaver coat requires 11–12 adult skins or 22–23 kits (local fur stores, personal communication, 2001).

Few European countries are ready to trap their recently reintroduced beavers for fur. In Germany it is illegal to possess a beaver skin. In Sweden, the population of the beavers reintroduced in the 1920s and 1930s grew by 20%–34% annually, so that hunting of beavers started in 1960 and continues today, as described earlier.

Time Line: Four Centuries of North American Fur Trade

15th Century

1497 John Cabot lands in North America.

16th Century

1535 Jacques Cartier's 2nd voyage to North America: St. Lawrence River, Quebec.

1536 Cartier visits Indian village Hochelaga ("Beaver Meadows"), today's Montreal.

Late 1500s Fishermen casually exchange manufactured goods for skins off Newfoundland.

17th Century

1608 Samuel de Champlain establishes fur-trading post at Statacona, today's Quebec City.

1609 Henry Hudson sails up the "Manhattan River," as far as Albany (New York).

1609 Champlain travels south to Lake Champlain.

1610 After sailing through Hudson's Strait, Hudson reaches south end of Hudson's Bay.

1611 Fur-trading post at Ville Marie, today's Montreal; 1641/42 settlement established.

1620 Pilgrim fathers found Plymouth Colony, a fur-trading post.

1627 Company of New France ("Company of the One Hundred Associates") incorporated under Richelieu. It exists until 1663.

1630 English force French back to St. Lawrence and Dutch back to Hudson River.

1635 Pynchon founds Springfield as fur-trading post.

1635 Concord, Massachusetts, founded as fur-trading post.

1654–61 The two *coureurs-de-bois* Radisson and Groseilliers explore southwest of Lake Superior. They reach Minnesota in 1659.

1670 Hudson's Bay Company is founded.

1682 The French "Compagnie du Nord" founded.

1686 French raid fur posts in James Bay.

1688, 1689 More fur raids occur.

1689 War between England and France officially declared ("King Williams War").

1690 More than 10,000 settlers in Canada.

18th Century

1713 "Beaver Wars" between England and France end with Peace of Utrecht. England gains Hudson's Bay, Newfoundland, and Nova Scotia (Acadia) from France.

1754 Youthful George Washington fails to dislodge the French from their fur post at Fort Duquesne (now Pittsburgh). An undeclared war leads to French and Indian War.

1756–63 French and Indian War (in Europe known as Seven-Years-War).

1759 British gain Quebec from French.

1763 Peace of Paris: Canada, Louisiana, and Cape Breton go from France to England.

1779 North West Company is founded, with headquarters in Montreal (founded anew in 1784).

1795 XY Company is founded.

1793–1814 Castorland Corporation owns land in New York. Later it becomes the summer home for Napoleon's brother.

19th Century

1803 Louisiana Purchase.

1804–06 Lewis and Clark Expedition, sent by Thomas Jefferson, to open path to the Pacific Ocean for American fur traders.

1809 John Jacob Astor (1763–1848) founds American Fur Company. He also founds Pacific Fur Company in 1810 and Southwest Company in 1811.

1811 Astor builds trading post at mouth of Columbia River, named Astoria.

1820 Kuhl describes and names the North American beaver scientifically: *Castor canadensis.*

1821 Hudson's Bay Company merges with North West Company.

1820s Policy of depleting all furbearers, to prevent settlement by Americans.

1823 Newhouse invents efficient steel trap.

1870 Hudson's Bay Company cedes Prince Rupert's Land (the Hudson's Bay watershed) to the government of Canada.

20th Century

1987 Hudson's Bay Company sells fur auction houses in Toronto and London.

1991 European Union bans fur imports from countries that use leghold traps. Ban is repeatedly postponed.

REFERENCES

1. Martin, C. 1978. Keepers of the game: Indian-animal relationships and the fur trade. Berkeley: University of California Press.
2. Schneider, P. 1997. The Adirondacks: a history of America's first wilderness. A John Macrae book. New York: Henry Holt.
3. Moloney, F. X. 1967 [1931]. Fur trade in New England 1620–1676. Hamden, Conn.: Archon Books.
4. Skinner, C. L. 1933. Beaver, kings, cabins. New York: Macmillan.
5. Newman, P. C. 1987. Canada's fur trading empire: three centuries of the Hudson's Bay Company. National Geographic Magazine 172: 192–228.
6. Hammond, L. 1993. Marketing wildlife: the Hudson's Bay Company and the Pacific Northwest, 1821–49. Forest and Conservation History 37: 14–25.
7. Newman, P. C. 1985. Company of adventurers. Volume 1. New York: Viking Penguin.
8. Burroughs, R. D., editor. 1961. The natural history of the Lewis and Clark Expedition. East Lansing: Michigan State University Press.

9. Ambrose, S. E. 1996. Undaunted courage: Merriwether Lewis, Thomas Jefferson, and the opening of the American West. New York: Simon and Schuster.
10. Newman, P. C. 1987. Company of adventurers. Volume 2. Caesars of the wilderness. New York: Penguin Books.
11. Guthrie, D., and J. Sedell. 1988. Primeval beaver stumped Oregon Coast trappers. Oregon State University Fisheries and Wildlife Department News and Views, June, p 14–16.
12. Lichatowich, J. 1999. Salmon without rivers. Washington, D.C.: Island Press.
13. Hartman, G. 1999. Beaver management and utilization in Scandinavia. In: P. E. Busher, and R. M. Dzięciołowski, editors. Beaver protection, management, and utilization in Europe and North America. New York: Kluwer Academic/Plenum. p 1–6.
14. Robinson, W. L., and E. G. Bolen. 1984. Wildlife ecology and management. New York: Macmillan.
15. Brooke, J. 2001. Portage la Prairie journal: beaver's dual role: national icon and huge pest. New York Times, March 31, 2001, p A4.

OTHER SOURCES

Devoto, B. ed. 1953. The journals of Lewis and Clark. Boston: Houghton Mifflin.

Sandoz, M. 1964. The beaver men: spearheads of empire. Lincoln: University of Nebraska Press.

Wishart, D. 1979. The fur trade of the American West 1807–1840. Lincoln: University of Nebraska Press.

Reintroductions and Other Transplants

18 chapter

Trading people for beavers: The Markgrave of Hesse is said to have wanted to replace his extirpated beavers so badly that in 1714 he offered soldiers in exchange. He traded tall young men ("Lange Kerls") to the "Old Dessauer," Prince Leopold I of Anhalt-Dessau, for beavers to transplant to his realm.

After H. Weinzierl, 1973

In 1904, at my request, the first Beaver appropriation bill ($500) was introduced in the New York Legislature. . . . That fall the Commission purchased 7 Beavers, 6 of which were successfully liberated the following spring. This year (1906) I obtained a second beaver appropriation ($1,000) from the Legislature, and the liberations will soon be resumed. A number of private citizens are co-operating with the State and liberating Beaver on their estates in the Adirondacks. The Adirondack Beaver supply is rapidly multiplying, and there are at present perhaps 50, as against 6 or 8 five years ago. Unquestionably, the Beaver restocking project is a complete success.

Harry V. Radford, in letter to E. T. Seton, 1906

Beavers of both species have been transplanted many times and in many parts of the world, both to reintroduce them where they had become extinct, and to introduce them as "exotics" to new areas.

Reintroductions

The purpose of most transplants has been to replenish depleted stocks or replace extirpated populations of the same species in parts of its former range. In northeastern North America beaver populations had been critically reduced or even extirpated in large areas at the end of the 19th century, so that several states in the United States and provinces in Canada took protective measures, particularly by prohibiting any further trapping. For instance, beavers in Massachusetts were extirpated in the early 1700s and reintroduced in 1928.[1] In 2002, Massachusetts counted about 70,000 beavers, after their population had tripled since 1996.[2] In another example, South Carolina beavers were extirpated by trapping and clearing of land for crops, and reintroduced in 1939.[3]

Most illuminating of the decline and recovery of North American beaver stocks is the history of the beaver in New York's Adirondacks. The original population in New York at the time of "the commencement of the white man's settlement" appeared to Harry V. Radford[4] to have been "not improbably several million." In 1671, according to a Dutch author, "fully 80,000 beavers a year" were provided by the Province of New Netherlands. By the year 1800 less than 5000 beavers were estimated to live in the entire Adirondack wilderness. This number further shrunk to 15 individuals around 1900. Already in 1869 Watson[5] had written about Essex County in the heart of the Adirondacks: "The beaver was found in great abundance throughout the region, by the first occupants. They no longer exist, it is believed, in the territory of Essex country."

In 1895 Radford and a group of conservation-minded sportsmen suggested successfully to the New York State legislature that it protect beavers completely from trapping. After strict protection was imposed, 20 beavers from the Province of Ontario were released in the Adirondacks between 1901 and 1906. Of these, 7 beavers were purchased from the Canadian exhibit at the Louisiana Purchase Exposition in St. Louis. They were brought to Old Forge in December 1904, and 6 were released in the Moose River's South Fork and the head of Big Moose Lake.[6] Today a marker in Old Forge commemorates the keeping of these beavers during the winter 1904/05 and their subsequent release (Plate 43). The state's Forest, Fish and Game Commission purchased 25 more beavers from Yellowstone National Park, and of these light-colored animals ("yellow beaver"), 14 were introduced to various parts of the Adirondacks.

The population proved resilient and recovered immediately. Bump and Cook[7] quoted Adirondack guides and trappers to have counted 40 beavers in 1905, 75 in 1906, and 100 in 1907. Residents complained about beaver damage by 1912, and in 1915 the Adirondack population was estimated to be at least 15,000 beavers. The many damage complaints around 1920 led to open beaver seasons in 1924 and 1925. During those two seasons, over 6000 beavers were trapped. The beavers were again completely protected until 1928 when 5000 were caught in one season.

In the 1920s and 1930s the New York Department of Environmental Conservation transplanted many "nuisance beavers" to many parts of southern New York, so that today only Long Island is free of beavers. In 1932, a "Beaver Bus" was put in service and two men worked full time to trap and transplant nuisance beavers.[7]

Today beavers have repopulated literally all suited habitats in New York State. Increasingly this species colonizes areas that are only marginal for beavers or that conflict with human land use.

The European beaver, *C. fiber*, had been extirpated from much of its range. Several countries have tried to reintroduce this species but met with varying success. In Switzerland, for instance, beavers were completely eliminated by the early 19th century. Hunting, rather than habitat destruction, led to its demise. Declared a cold-blooded animal or fish, it was eaten during Lent; its pelt was used for beaver hats; and castoreum was much sought after as it was considered a panacea. Furthermore, the beaver was persecuted because it allegedly competed with humans for fish. River regulation and habitat destruction by development occurred only after the beaver was already extinct in Switzerland. Starting in 1956, beavers were released in several regions. Only three areas proved to be suited for beavers; the introduced animals once again abandoned most other areas. Developed riverbanks and fluctuating water levels deprived beavers of year-round underwater access to their lodges. Some streams are too steep, and lakes must have quiet bays protected from waves. Finally, undesirable plant species render many areas unattractive.[8] Stocker[8] pointed out that the Swiss landscape has been so radically changed in the 150 years since beavers were eradicated, that the remaining beaver habitat represents little more than isolated pockets in an extremely developed country. He takes a pessimistic view of opportunities for dispersal and colonization of new areas and recommends a guarded and careful review of any further attempts at reintroduction into Switzerland.

To the north, Bavaria has been more successful. The last beavers in Bavaria had been observed in 1867,[9] and reintroductions started in 1966. In 1996, the beavers had spread to many areas: 21.4% of the topographic maps contained beaver locations.[10]

In the former East Germany, reintroduction have been very successful. There, after near extinction, special protective measures and transplants had increased the beaver population of 90 families in 1952 to 380 families in 1982[11] and 430 families by early 1986. Today they have spread and connected to other river systems.

Finland lost its beavers in 1868 and Sweden, in 1871. These countries started reintroductions in 1935 and 1922, respectively. About 90% of the current Finnish beaver population belongs to the North American species. It took about 40 years until management of the still-growing populations became necessary.[12] In Russia where larger numbers were released simultaneously, this point was reached in only 20 years.[13]

Poland presents a success story. The watershed of the Vistula, entirely within, and comprising the larger part of Poland, lost its beavers in the early 19th century. Starting in the 1940s, and again between 1975 and 1985, when 168 beavers were released, wildlife managers restored beaver populations along the entire Vistula basin, from the Masurian and Pomeranian Lakelands in the north of the country to the Carpathian Mountains in the south. The animals came from the wild in the Baltic region of Russia, or from a beaver farm. On average, four pairs were released 2–20 km from each other. Such "populations" of 4 pairs were never more than 100 km apart from each other. This ensured that young dispersing animals would find partners to establish new colonies. Many introduced beavers stayed in developed areas but suffered high mortality there. They preferred streams to lakes and chose sites with high banks to dig burrows. The offspring derived from the introduced animals dispersed up to 20 km, except in drought years, when they settled only 30–100 m from their parental colonies. The number of active beaver sites increased by 20% annually.[13]

Movements of Transplanted Beavers

Translocated beavers often don't stay where they have been released. They may move great distances, although not necessarily into the direction of the place where they were caught. Some beavers stay at the release site. Therefore, the distances of their movements range from zero to the record of 390 km as the water flows, estimated in Massachusetts.[14] Table 18.1 lists some distances traveled by transplanted beavers in different regions of North America and in Russia.

Beavers as Exotic Introductions

New locales for introductions beyond the natural range of either beaver species include such distant areas as Tierra del Fuego in Argentina (Plate 44), Aus-

Table 18.1 | Some Distances Traveled by Translocated Beavers from Their Release Site

Geographical Area	Average Distance (km)	Maximum Distance (km)
James Bay, Canada[15]	18 A	66 A
Goose Bay, Labrador[16]		120 (one animal) A
Maine[17]	11.3 S	25.8 S
Wisconsin[18]	7.4 A	48 A
Massachusetts[14]	—	390 S
North Dakota[19]	14.6 S 9.5 A	—
Colorado[20]	16.7 S	48 S
New Mexico[21]	12.9 S, male 9.7 S, female	—
Volga-Kama Reserve, Russia[22]	4–16 (upstream)	—

Note: A, air distance; S, distance along stream.

tria, and Finland. In Tierra del Fuego, 25 pairs of the North American beaver were introduced at Lago Fagnano in 1946. Since then the habitat has become saturated with beavers. The only woody species serving as food and building material is southern beech, locally known as *lenga* (*Nothofagus pumilio*). The beaver ponds have created new habitat for the spectacled duck (*Anas specularis*) and speckled teal (*A. flavirostris*).[23]

To quote another example, two pairs of North American beavers were introduced into Finland in 1937. In one area alone (South Savo Game Management District), the beaver population increased from 1100 to 3000 individuals in 400 and 1280 colonies, respectively, during the years 1983–97.[24]

REFERENCES

1. Shaw, S. P. 1984. The beaver in Massachusetts. Massachusetts Department of Conservation Research Bulletin 11.
2. Sterba, J. P. 2002. As forest reclaims American East, it's man vs. beast. The Wall Street Journal, May 21, 2002, p A13.
3. Woodward, D. K., R. B. Hazel, and B. P. Gaffney. 1985. Economic and environmental impacts of beavers in North Carolina. In: Proceedings of the Second Eastern Wildlife Damage Control Conference; Raleigh, North Carolina. Raleigh: North Carolina State University. p 89–96.
4. Radford, H. V. 1907. History of the Adirondack beaver. Annual reports of the Forest, Fish and Game Commission of the state of New York for 1904–1905–1906. p 389–418. Albany, N.Y.: The Forest, Fish, and Game Commission.
5. Watson, W. C. 1869. The military and civil history of the county of Essex, New York. Albany, N.Y.: J. Munsell.
6. Decker, D. J. 1980. The beaver: New York's empire builder. Conservationist 35(3): 15–17, 40.
7. Bump, G., and A. H. Cook. 1941. Black gold: the story of the beaver in New York State. Conservation Department of New York State. Management Bulletin 2.
8. Stocker, G. 1985. Die Wiedereinbürgerung des Bibers in der Schweiz. Der Schweizer Förster 10: 477–490.
9. Weinzierl, H. 1973. Projekt Biber. Kosmos-Bibliothek. Volume 279. Stuttgart: Franckh'sche Verlagshandlung.
10. Zahner, V. 1996. Der Einfluss des Bibers (*Castor fiber*) auf gewässernahe Wälder [doctoral dissertation]. Munich: Munich University.
11. Heidecke, D. 1986. Bestandssituation und Schutz von *Castor fiber albicus* (Mammalia, Rodentia, Castoridae). Zoologische Abhandlungen, Staatliches Museum für Tierkunde, Dresden.
12. Hartman, G. 1999. Beaver management and utilization in Scandinavia. In: P. E. Busher, and R. M. Dzięciołowski, editors. Beaver protection, management, and utilization in Europe and North America. New York: Kluwer Academic/Plenum. p 1–6.

13. Zurowski, W., and B. Kasperczyk. 1988. Effects of reintroduction of European beaver in the lowlands of the Vistula basin. Acta Theriologica 33: 325–338.
14. Hodgdon, H. E. 1978. Social dynamics and behavior within an unexploited (*Castor canadensis*) population [Ph.D. dissertation]. Amherst: University of Massachusetts.
15. Courcelles, R., and R. Nault. 1983. Beaver programs in the James Bay area, Quebec, Canada. Acta Zoologica Fennica 174: 129–131.
16. Chubbs, T. E., and F. R. Phillips. 1994. Long distance movement of a transplanted beaver, *Castor canadensis*, in Labrador. Canadian Field-Naturalist 108: 366.
17. Hodgdon, K. W., and J. H. Hunt. 1953. Beaver management in Maine. Bulletin 3. Maine Department of Inland Fisheries and Game, Game Division. Augusta, Maine: Department of Inland Fisheries and Game.
18. Knudsen, G. J., and J. B. Hale. 1965. Movements of transplanted beavers in Wisconsin. Journal of Wildlife Management 29: 685–688.
19. Hibbard, E. A. 1958. Movements of beaver transplanted in North Dakota. Journal of Wildlife Management 22: 209–211.
20. Denny, R. N. 1952. A summary of North American beaver management, 1936–1948. Current Report 28. Colorado Game and Fish Department. Denver: Colorado Fish and Game Department.
21. Berghofer, C. B. 1961. Movement of beaver. Proceedings of the Annual Conference of the Western Association of State Game and Fish Commissioners 41: 181–184.
22. Gorshkov, Y. A., A. L. Easter-Pilcher, B. K. Pilcher, and D. Gorshkov. 1999. Ecological restoration by harnessing the beavers. In: P. E. Busher and R. M. Dzięciołowski, editors. Beaver protection, management, and utilization in Europe and North America. New York: Kluwer Academic/Plenum. p 67–76.
23. Dietrich, U. 1985. Beobachtungen an Kanadabibern *Castor canadensis* in einem Einbürgerungsgebiet auf der Insel Feuerland, Südamerika. Säugetierkundliche Mitteilungen 32: 241–244.
24. Härkönen, S. 1999. Management of the North American beaver (*Castor canadensis*) on the South-Savo game management district, Finland (1983–1997). In: P. E. Busher and R. M. Dzięciołowski, editors. Beaver protection, management, and utilization in Europe and North America. New York: Kluwer Academic/Plenum. p 7–14.

"Nuisance Beavers" Claim Their Land

chapter 19

> Possibly the most serious damage is that caused to timber by flooding, of which Johnson has made a study in the Adirondacks, and this in spite of the complaints which were made, does not appear to be as serious as some of those opposed to the beavers claimed. It so happens that my own observations show comparatively little of this damage. While I have seen many trees thus killed, by far the greater portion have been aspens, alders and willows, which are not considered of any particular value. The commercial value of the pines, firs and spruces which I have seen destroyed by beaver ponds would be small, though they sometimes made a rather conspicuous showing when standing in a pond with their brown tops contrasting with the green living trees about them.
>
> *Edward R. Warren, 1927*

What are "Nuisance Beavers"?

Beavers can modify the landscape in dramatic ways that are often unwelcome to humans. "Nuisance beavers" attract attention by the media. In the spring of 1999, beavers appeared at the tidal basin in the nation's capital and cut down several ornamental cherry trees. They had to be trapped alive and transferred to an undisclosed destination. In the other North American capital, Ottawa, beavers have felled cottonwood trees along the scenic Ottawa River Parkway, used for recreational walking.[1]

One day in the summer of 1986 a local conservation officer in Delaware County, New York, received a call from then U.S. Senator Daniel Patrick Moynihan. Beavers had taken up residence in a stream on his farm. There are now so many landowners' complaints about beavers and no places left to transfer them to that the game warden could not help. In the state of New York, "Licensed Nuisance Agents" now destroy such nuisance beavers.[2] In Massachusetts, "wildlife-damage control professionals" charge $150 to remove a "problem beaver," and $750 for a family of 5 beavers.[3] The profession is organized into the "National Wildlife Control Operators Association" with about 500 members. *Wildlife Control Technology* is the trade magazine.

Nuisance beaver is a misnomer, an anthropocentric term at best. Damage to crops, tree plantations, roads, water supplies, and recreational facilities such as golf courses and campgrounds results from human encroachment on the beaver's habitat, not the other way around. In short, the beaver's recent spectacular expansion of its range southward from wilderness to developed land has intensified the conflict between beast and man. Almost everywhere, humans' and wildlife's uses of the landscape overlap.

Nevertheless, the beaver causes economic losses by flooding and softening roads (Plate 45) and flooding often tens of acres of farmland, as in Fulton County, New York, and golf courses, as near Acadia National Park in Maine. Aggravating enough, such damage is dwarfed by matters of life or death. In Canada, a train derailed on a rail bed softened by beaver flooding between Ottawa and James Bay. A person lost his leg and sued the railroad for a large amount. The lawsuit revolved around the question of who is responsible: the railroad who owns a 150-foot right-of-way, or the Department of Natural Resources that has jurisdiction beyond (M. Lavendière, personal communication, 1985). A passenger train experienced an even more serious accident. On July 7, 1984, an Amtrak train traveling between Washington, D.C., and Montreal derailed in Vermont while crossing a creek. The National Transportation Safety Board concluded later that a series of beaver dams upstream had collapsed after heavy rains, causing a flash flood that washed away the rail tracks' embankments at the stream crossing. Seven cars and both locomotives derailed, 5 people died, and 26 were seriously injured.[4] Thus, the enormous masses of water often impounded by beavers may prove a time bomb if released at once through one or several broken dams. Figure 19.1 shows tracks of the Old Adirondack Railroad between Old Forge and Lake Placid, now suspended in the air, after a series of beaver dams broke and washed out the railway embankment.

On agricultural land near rivers beavers often consume and damage crops such as corn.[5] In the American South, beavers damage flotation blocks under boat docks and marinas.[6] In a survey in North Carolina, girdled and flooded timber, blocked culverts, and flooded crops led the list of complaints by property owners. Three times as many owners reported damages than benefits, such as waterfowl hunting on beaver sites, aesthetic enjoyment, and fishing.[7] To save the endangered Paiute cutthroat trout in California, beavers that were changing its habitat had to be removed.[8] California also suffered losses of fruit and nut trees, sugar beets, and other crops, but above all softening and breaking of the levees designed to retain water in the Sacramento–San Joaquin Delta.[9] In nonimpounded bottomland forests in Mississippi, beavers girdle or fell economically important trees, particularly spruce pine, loblolly pine, and sweetgum.[10]

Beavers at "nuisance sites" are younger, 2–4 years old, and live in smaller groups (average: 3.7/colony) compared to other, established family colonies.

Figure 19.1 | Tracks of the Old Adirondack Railroad suspended high in the air after beaver dams broke and washed out the railway embankment.

Young, colonizing beavers have small litters. It is recommended to trap these beavers annually before they are firmly entrenched and cause unbearable damage. Government or contract trappers are essential in this effort. Because nuisance beavers are young and their colonies are small, the pelts may not be in their prime at the time of the nuisance complaint and are worth little. Therefore, fur trappers do not have the motivation or time to completely remove the whole colony.[11]

In more densely populated Europe, the beaver, *C. fiber*, causes damage much sooner. As soon as it leaves its few restricted sanctuaries in the narrow strips along the stream banks, conflicts may arise. In Bavaria, for instance, beavers feed in sugar beet fields and undermine stream banks.[12]

Beaver and Acid Waters: Problem or Solution?

The sulfur dioxide and nitrous oxides that emanate from power plants, factories, automobiles, and heated homes combine with water in the air or on the ground to become sulfates and nitrates, respectively. The continuing influx of sulfates and nitrates via rain, snow, hail, fog, and running and seeping water on the ground acidifies lakes and ponds (i.e., their pH decreases).

Fortunately, many bodies of water have the capacity to counteract this acidification by having available sources of alkalinity. Until now the primary source of such alkalinity has been thought to be geochemical processes, such as weathering of rocks in the watershed above the lake or pond. Recently, however, it was noticed that the water in some lakes in northwestern Ontario had higher alkalinities than their inflow streams.[13] This suggested that processes in the lake itself increase alkalinity. This increase can be accomplished either by a higher concentration of cations, such as calcium and potassium that diffuse from the sediment into the water column, or by removal of anions, such as sulfate and nitrate. In the lakes investigated by Schindler and coworkers,[13] three processes accounted for most of the generation of alkalinity: sulfate removal by sulfate reduction, followed by iron sulfide precipitation or binding as organic sulfur; exchange of H^+ for calcium ions in the sediment; and nitrate reduction. These alkalinity-increasing processes take place in the upper, bacteria-rich layer of the bottom sediment.

Sediments are formed particularly in slow-flowing bodies of water with decaying vegetation. A beaver pond is a perfect example. So on the one hand, the beaver creates the conditions for biological sulfate and nitrate reduction to take place. On the other hand, organic matter accumulates in the pond. Organic carbon, nitrogen gas, and sulfur will be oxidized to carbon dioxide, nitrate, and sulfate respectively. These all acidify the water/soil system. Whether the beaver can be our ally in the fight against acid precipitation to any significant degree, or contributes more to acidification, remains to be seen.

REFERENCES

1. Brooke, J. 2001. Portage la Prairie journal: beavers' dual role: national icon and huge pest. New York Times, March 31, 2001, p A4.
2. King, W., and W. Weaver Jr. 1986. Beavers. New York Times, August 15, 1986, p A10.
3. Sterba, J. P. 2002. As forest reclaims American East, it's man vs. beast. The Wall Street Journal, May 21, 2002, p A13.
4. Associated Press. 1985. Amtrak accident laid to beavers. New York Times, December 14, 1985.
5. Dieter, C. D., and T. R. McCabe. 1988. Beaver crop depredation in eastern South Dakota. Prairie Naturalist 20: 143–146.
6. Wade, D. A. 1987. Economics of wildlife production and damage control on private lands. In: D. J. Decker and G. R. Goff, editors. Valuing wildlife. Boulder, Colo.: Westview. p 154–163.
7. Woodward, D. K., R. B. Hazel, and B. P. Gaffney. 1985. Economic and environmental impacts of beavers in North Carolina. In: Proceedings of the Second Eastern Wildlife Damage Control Conference; Raleigh, North Carolina. Raleigh: North Carolina State University. p 89–96.
8. Hunter, H. C. 1976. Proposed beaver removal in Cottonwood Creek. Environmental

Analysis Report, U.S. Forest Service, Inyo National Forest, White Mountain Ranger District, Calif. May 7, 1976.

9. Fitzgerald, W. S., and R. A. Thompson. 1988. Problems associated with beaver in stream or floodway management. In: A. C. Crabb and R. E. Marsh, editors. Proceedings of the Thirteenth Vertebrate Pest Conference, Davis, California: University of California, Davis. p 190–195.
10. Bullock, J. F. 1982. The ecological and economic impact of beaver (*Castor canadensis*) on bottomland forest ecosystems of Mississippi [M.S. thesis]. Mississippi State: Mississippi State University.
11. Peterson, R. P., and N. F. Payne. 1986. Productivity, size, age, and sex structure of nuisance beaver colonies in Wisconsin. Journal of Wildlife Management 50: 265–268.
12. Schwab, G., W. Dietzen, and G. V. Lossow. 1994. Biber in Bayern—Entwicklung eines Gesamtkonzeptes zum Schutz des Bibers. Beitraege zum Artenschutz 18: 9–44. Munich: Bayerisches Landesamt fuer Umweltschutz.
13. Schindler, D. W., M. A. Turner, M. P. Stainton, and G. A. Linsey. 1986. Natural sources of acid neutralizing capacity in low alkalinity lakes of the precambrian shield. Science 232: 844–847.

chapter 20

Needed: An Ecosystems Engineer for Habitat Restoration and Other Services

A beaver pond is a leaky reservoir, a kind of spring as it were, and if stored full during rainy days the leakage from it will help maintain stream-flow below during the dry weather. Beaver works thus tend to distribute to streams a moderate quantity of water each day. In other words, they spread out or distribute the water of the few rainy days through all the days of the year.

Enos A. Mills, 1913

Ecosystem Services

The beaver, untiring dam builder, can and should be our important ally in wetland restoration. For millennia, beavers have created and maintained wetlands that stored water and kept the water table high. Their leaky dams evened out the flow of streams. For instance, a side canyon (Butler Wash) of the San Juan River in southern Utah is said to never flood, thanks to beaver dams.[1] The two branches of the Satsop River in Washington State differ in their water flow after rainstorms and snow thaws: The East Fork runs through more flat ground and less elevation gradient. It has more wetlands and beavers than the "Middle Fork" and rarely floods. By contrast, the steeper Middle Fork with fewer beavers floods easily (G. Schirato, Washington Wildlife Program, personal communication, 2002).

Leaky dams provide an ideal compromise of "ecosystems services." First, they control flooding better than solid concrete dams because of their gradual release of water. Second, they slow down the flow sufficiently to allow time for cleansing the water. According to a computer model, water flowing through a 1-square-mile area (2.59 km^2) with no dams resides for only 3–4 hours, while the same area with a 5-foot-high (1–5 m) leaky dam retains water for about 11 days. Tight dams retard water almost twice as long (19 days). Long retention times for water behind dams permit the removal of toxins and excess nutrients: to clear the water of nitrogen, phosphorus, or herbicides, such as atrazine, requires a retention time of 6–8 days in the water body. Three processes accomplish this: deposition (sedimentation), microbial decomposition, and chemical transformation, augmented by some filtering (D. Hey, personal communication, 2002).

Over the centuries, humans have drained wetlands and hunted beavers to extinction in many areas. This loss of beavers and wetlands has worsened floods

Table 20.1 | Differences between Intact and Degraded Streams in Arid Area

	Intact or Restored Stream	Degraded Stream
Productivity	5000 lb/acre	200 lb/acre
Silt loss (tons/day/5miles)	4	109
Depth of woody vegetation (feet from water)	30–40	Only at water's edge

and erosion and degraded habitats because it lowered the water table. During droughts, eroded streams with their low water level drain water out of the soil of the surrounding watershed. Overgrazed, beaverless stream valleys in arid settings suffer severe habitat degradation, as Table 20.1 shows.[2,3] Degraded streams produce less biomass along their banks, which in turn lose much soil.

If beavers could be induced to settle along streams in degraded areas, they might fix the problems by impounding water and raising the water table. Exactly that has been undertaken in several places, serving as models for future bioremediation projects. For best results, whole families should be released.

In 1977, beavers on Horse Creek in southwestern Wyoming moved into a degraded area and built a dam from sagebrush and rabbitbrush, for lack of other woody material, especially trees. Each spring, these dams washed out. Bruce Smith, then with the Bureau of Land Management, provided aspen branches, and the beavers immediately went to work. Within days, they had incorporated the aspen into their dams. At other creeks beavers also built dams from materials provided by wildlife managers, and thus stabilized the stream banks.

The second phase of this restoration effort was different. Wildlife experts delivered not only aspen but also the beavers to eroded stream sections in need of restoration on Currant Creek. These newcomers built three major dam complexes and raised the water table by 0.3–1.0 m.[2–4] At the end of the third year after reintroduction, willow had grown 1.6–2.0 m in height where the beavers had raised the water table. By lifting the water level, the beavers accelerated a recovery that would be much slower with rest from livestock grazing alone. The restored stream carried much less sediment with it. The different stretches of Currant Creek illustrate this effect: It carried 34 ounces (about 1 kg) of sediment from public land in every 100 cubic feet (2.8 m^3) of water. After it had passed 5 miles of private land with beaver dams and intact riparian vegetation, the stream held 90% fewer solids. Below that, half a mile of overgrazed pasture with degraded bank vegetation reversed this pattern: the particle content increased again by 110%. Finally, the level reached 112 ounces/100 cubic feet of water (about 3.5 kg/2.8 m^3) on 2 more miles of degraded riparian habitat.[4] Nitrogen and phosphorus adsorb to clay. Therefore, the sediment in the beaver ponds also trapped nitrogen and phosphorus. The ponds serve as sinks for these nutrients that otherwise stimulate the growth of algae and other water plants, a process known as *eutrophification.*

At one stream (Sage Creek), beavers tried to establish themselves, but their dams washed out every spring. The managers strung tires on a steel cable, or net wire along the small beaver dams. The animals built 60 cm on top of that, and these larger dams held for several seasons. Repeated beaver releases and provisioning with aspen at other creeks resulted in dam building and established colonies.[2–4] The dams slow down the water flow, reducing erosion and enhancing the deposition of sediment behind the dam. An average dam there can retain 7000 cubic yards (5355 m^3) of sediment.[5] The sediment permits the growth of shrubs and trees. Sage Creek ultimately had five tire-reinforced dams over about 800 m (half mile). The project cost $4000 for materials, plus the labor by federal employees. Human engineers would have charged $100,000 or more for erosion control structures with the same effect.[6] The effects of beaver dams on the streams are listed in Table 20.2.

Similar releases of beavers for revitalization of stream banks have been undertaken in Utah (Echo Creek), Montana (Howard Creek), and Idaho (Cooper Creek; Plates 46 and 47).[7] At Cooper Creek and Grant Creek, the latter a tributary to the Big Lost River in the Pioneer Mountains, Idaho, beavers affected the water more than just by impounding it behind a dam. Soil cores taken 100 feet (33 m) from the creek showed that the water table in the surrounding area had risen to the level of the water in the pond. Thus, beavers store much more water than meets the eye. Second, such a reservoir of soil water moderates water temperatures in the ponds and the creek. The beaver pond and the surrounding soil exchange water: At low stream levels, water seeps back into the creek, and at high levels, water spreads in the opposite direction, through the soil of the floodplain. In winter, soil water percolating back warms the stream, while in summer cool groundwater oozing back into the creek retards warming of the pond and stream

Table 20.2 | Effects of Beaver Dams on Streams

Effects
Trapped sediment
Reduced stream velocity
Elevated water table
Attenuated seasonal water table fluctuation
Favored growth of willow and other riparian plants
Stabilized banks
Reduced sediment transport by 90%
Sediments in pond trapped phosphorus and nitrogen
Thus improved water quality

Source: Reference 4.

water. Particularly in winter, the warming effect of the groundwater may make the difference for survival of fish. In addition, the deeper water of a beaver pond permits fish to choose from different layers. Storage of large volumes of water is particularly important in arid climate streams. In summer, Cooper and Grant Creeks flow at a rate of only 0.6–1.0 cubic feet/second (270–450 gallons/minute, or 1200–2045 liters/minute) (Lew Pence, personal communication, 2002).

The Zuni Indian Reservation in New Mexico has released beavers into degraded watersheds. Within 1–2 weeks after reintroduction, these animals built dams that slowed the water flow and permitted the sediment to settle, thus raising the streambed. As a consequence, the water spread over a wider area. Larger pools persisted year-round. Much more riparian vegetation grew up, and fish, amphibians, birds, deer, and elk moved into this improved habitat. Specifically, a federally endangered bird species, the southwestern willow flycatcher (*Empidonax traillii extimus*), found a home in the restored willow and cottonwood vegetation. In addition, the raised water table killed the salt cedar (*Tamarix petandra*), an aggressive, unwanted invader.[8]

Such systems of restored vegetation not only are cheap but also become self-supporting after only 2–3 years.

The Eurasian beaver is helping to restore streams in the Volga-Kama National Preserve in Tatarstan, Russia. Upstream agricultural regions discharge sediment that fills up lakes and bogs. As a result, the landscape dries up and loses certain plant and animal species. Starting in 1996, beavers were released into lakes. They moved 4–16 km upstream, settled along streams, and built dams. It is hoped that these impoundments catch the troublesome sediment before it reaches a lake or bog.[9]

Beavers have been "rediscovered" as environmental engineers and ameliorators every few decades. In the 1930s and 1940s, they were released in the West to control stream flow and soil erosion. The Forest Service even parachuted beavers into remote forest meadows to build dams for reservoirs for fighting forest fires.[10]

Large-Scale Remediation

Donald Hey and Nancy Philippi[11] of the Wetlands Initiative have proposed a much more ambitious restoration scheme. Over the 20th century, flooding of the Mississippi-Missouri-Ohio Rivers system has become more frequent and more costly. Mean annual flood damage in the upper Mississippi River Basin has increased 140%, in 1993 dollars. Levees, constructed for protection from floods, actually exacerbate the problem. They increase river stage and velocity.

During the 19th century, vast areas of wetlands had been drained. Countless beaver ponds and marshes that used to retain water were eliminated. In 1600, beaver ponds covered 11% of the upper Mississippi and Missouri Rivers' watershed above Thebes, Illinois. In 1990, only 1% of these, or 0.1% of the watershed, remained. The 1993 flood in the Midwest was an inevitable aftermath of this wetland reduction. Before humans modified the landscape, vegetation, organic soils, and wetlands captured precipitation where it fell. Runoff found its way into creeks, where beaver ponds stored water. After torrential rains and the spring snowmelt, the excess water spread out over wetlands. Along the larger rivers, floodplains accommodated the overflow. The water slowed down, evaporated, percolated into the soil, or slowly receded into the river channel. Unfortunately, settlers and their descendants cleared the land, straightened out the streams and rivers, and paved over large areas. More than half of the beaver ponds and marshes in North America have been destroyed. Agriculture and settlements claimed the floodplains, and levees constrained the rivers. Seventy percent of the water-storing capacity of the soils disappeared during the past 150 years. Designed to retain floodwater, the levees actually often fail. One thousand of them did so along the upper Mississippi in 1993.

Hey and Philippi calculate that the excess water that passed St. Louis during the 1993 flood would have covered 13 million acres (52,600 km^2), standing 3 feet (90 cm) deep, as in a marsh. This is half of the wetland acreage drained since 1780 in the upper Mississippi River Basin. The authors[11] suggest restoration of about 13 million acres of the original vegetation and wetlands in the watershed. A prairie-forest matrix could intercept water, while strategically placed wetlands could cushion water levels. The upper 18–24 inches (46–61 cm) of the soils should be enriched with organic matter. Calculations show that as little as 3% of the watershed needs to be restored to absorb the water volume of the devastating 1993 floods. A 100-year flood flow (1,100,000 cubic feet/second or about 31,000 m^3) would require 13.3 millions of acres (about 5.4 million hectares) in wetlands, or 2.9% of the watershed investigated by Hey and Philippi. Recovered wetlands would also store agricultural runoff and urban wastewater for treatment.

Local Solutions

Municipalities and farms can benefit from the beaver's water-storing skills. Beaver ponds provide good sources of water for fire protection and livestock. A case in point is the Village of Sharon Springs, New York. For years, two beaver ponds have served as water tanks for the town's fire protection, and as of this writing, still do (D. Barlow, Town Clerk, personal communication, 2001).

During a drought from 1987 to 1989, beavers served Grant City, Missouri, very well. Just when the Grand River went dry, and water for corn, soybeans, and livestock became scarce, these rodents built seven dams and created pools at a stretch of the river that feeds the town's wells.[12]

Wetlands have been celebrated as the "kidneys of the landscape." Plants and organisms of the rhizosphere (root layer) can cleanse water of pollutants. Plants in particular can deal much better with low levels of dispersed but harmful pollutants than other methods. For instance, in the U.S. state of Georgia, the aquatic weed parrot feather in artificial wetlands serves to destroy trinitrotoluene, found on military firing ranges. Wetlands also clean up polluted water from landfills, slaughterhouses, cider mills, sewage plants, fish farms, and parking lots. Willows and poplars are being used for such "phytoremediation." Steven A. Rock, an engineer with the National Risk Management Research Laboratory of the U.S. Environmental Protection Agency in Cincinnati, speaks of a "self-assembling solar-powered pump-and-treat-system."[13] Needless to say, management of water quality looms as an ever more important task of the 21st century. Wetland conservation and restoration play a key role in this effort, and we should recruit the beaver, the greatest original wetland conservationist, for this effort, wherever possible.

REFERENCES

1. Aton, J. M., and R. S. McPherson. 2000. River flowing from the sunrise. Logan: Utah State University Press.
2. Apple, L. L., B. H. Smith, J. D. Dunder, and B. W. Baker. 1995. The use of beavers for riparian/aquatic habitat restoration of cold desert, gully-cut stream systems in southwestern Wyoming. In: G. Pilleri, editor. Investigations on beavers. Volume 4. Berne: Brain Anatomy Institute. p 123–130.
3. Olson, R., and W. A. Hubert. 1994. Beaver: water resources and riparian habitat manager. Laramie: University of Wyoming.
4. Apple, L. L. 1985. Riparian habitat restoration and beavers. In: Riparian ecosystems and their management. United States Department of Agriculture Forest Service General Technical Reports. RM-120. Washington, D.C.: U.S. Department of Agriculture. p 489–490.
5. Naiman, R. J., J. M. Melillo, and J. E. Hobbie. 1986. Ecosystem alteration of boreal forest streams by beaver (*Castor canadensis*). Ecology 67: 1254–1269.
6. Johnson, P. 1984. The dam builder is at it again! National Wildlife 22: 8–15.
7. Pence, L. 1999. Beaver: a tool for riparian management. Presentation at workshop "Beaver and Common-Sense Conflict Solutions," Estes Park, Colorado, September 7–9.
8. Albert, S., and T. Trimble. 2000. Beavers are partners in riparian restoration on the Zuni Indian Reservation. Ecological Restoration 18: 87–92.
9. Gorshkov, Y., A. Easter-Pilcher, B. Pilcher, and D. Gorshkov. 1999. Ecological restoration by harnessing the work of beaver. In: P. E. Busher and R. M. Dzięciołowski, editors. Beaver protection, management, and utilization in Europe and North America. New York: Kluwer Academic/Plenum. p 67–76.
10. Bergstrom, D. 1985. Beavers: biologists "rediscover" a natural resource. Forestry Research West. U.S. Department of Agriculture Forest Service.

11. Hey, D. L., and N. S. Philippi. 1995. Flood reduction through wetland restoration: the upper Mississippi River Basin as a case history. Restoration Ecology 3: 4–17.
12. Robbins, W. 1989. Town thanks beavers for water in drought. New York Times, July 12, 1989.
13. Revkin, A. C. 2001. New pollution tool: toxic avengers with leaves. New York Times, March 6, 2001. p F1.

Conservation and Proactive Management

21 chapter

> From the esthetic standpoint, it would seem scarcely advisable to allow the beaver to cut all the large aspen bordering the [Yellowstone] Park streams and ponds or to flood every convenient spruce flat. . . . Unless a natural balance among the Park animals can be maintained, and the beaver population kept at a constant normal figure, we may expect recurrent cycles of gradual stocking with aspen and beaver followed by rapid exhaustion of the food supply and dispersal of these animals elsewhere.
>
> *E. R. Warren, 1926*

Why Beaver Management?

Many urban people ask the question, Should we humans manage beaver populations? The short answer is that we have no choice. Where beavers are rare, reintroductions and the conservation of beavers and the restoration of their habitat are called for. Burgeoning beaver populations, on the other hand, require proactive management to prevent adverse consequences such as flooding and damage to trees. In either case, conservation of functioning wetlands constitutes the main task. All too often, beavers cause some damage to unprepared property owners, and immediate action is called for. Instead of having to put out fires, we should plan ahead. Proactive management avoids conflicts between wildlife and people and between different stakeholders. In less inhabited regions, beavers are managed for a sustainable fur harvest, since predators no longer regulate their numbers. Here again, a balance between a healthy beaver population and its intact habitat is the stated goal.

In North America, indigenous people have used beavers as food and medicine and for clothing since time immemorial. They managed beavers in the sense that they did not overharvest them. They enjoyed a sustainable yield. First, rampant trapping by Europeans and their Native American suppliers, without conservation or management, and later habitat change—particularly draining of wetlands for development and agriculture—extirpated the North American beaver over large parts of its range. In Europe, draining of wetlands for agriculture, straightening of rivers, and elimination of floodplains, as well as the ever increasing settling of people in former beaver habitat, took place even earlier.

Ever since, people and beavers have coexisted in an unstable way. At many different times and places, beavers were deemed either too scarce or too numerous. The first condition led us to reintroduce beavers and protect these animals and their habitat, while the latter compelled wildlife managers and property owners to control beavers by shooting, trapping, or translocation.

At the beginning of the 20th century, large areas of the more developed American Northeast lacked beavers. During the first decade, they were reintroduced, for instance, in 1905 in the Adirondack Mountains of New York State. Chapter 18 on transplants lists more examples. Nursed along for the next decades, beavers became numerous again in the middle of the century. By now humans had not only intruded into beaver habitat but also profoundly altered it. Golf courses, shopping centers, housing developments, and sewage plants had sprouted into former wetlands. Young beavers traveling the waterways in search of a home would literally dig in at now marginal places and be promptly declared "nuisance beavers." Where humans had intruded into beaver habitat, the returning beavers now were perceived to intrude on human habitat. Many roads and bridges that experience flooding as a result of beavers today were constructed at a time when beavers were absent from that particular area. Homeowners and landowners frequently call their wildlife agency to remove the beavers. The state fish and game departments have become strained for resources to deal with the problems of flooding and tree cutting by beavers. An increasingly urban and suburban population resists shooting and trapping to kill. For a while, live-trapping and transplanting the animals to more suitable beaver habitat worked as an acceptable alternative. But now managers cannot find new homes for the so-called nuisance beavers because most landowners who welcome beavers have them already. In the 21st century, management of burgeoning beaver populations requires new approaches. Manitoba, for instance, pays a bounty of $15 for each beaver, with few questions asked. Considering the depressed fur prices and the work required to prepare a pelt, such "nuisance trapping" can be attractive.

How Large Should We Allow the Beaver Populations to Be?

Population density should be low enough to allow young beavers leaving their parental colony to find places to settle, without becoming nuisance beavers that interfere with human land use. Left to their own devices, beaver populations will grow to capacity. There is some self-regulation of populations: sparse populations produce more offspring than saturated populations, but beavers will be in conflict with humans long before that effect sets in. Therefore, beavers need enough vacant suitable sites that are available to dispersing youngsters for colonizing. Prudent management provides such immigration sites by keeping enough stream sections and other wetland areas free of beavers. On a landscape scale, North American wildlife managers keep beaver populations low enough that dispersing youngsters can find suitable habitat without engendering conflicts with humans. They aim at 10%–30% occupancy of potential beaver sites. Agencies in different states or provinces have different target levels. Fur trapping by licensed private trappers seeks to achieve that goal. It can also be accomplished to some extent by live-trapping and transplantation of beavers. Unfortunately, often there are not enough sites where the beavers could be released. Fur trapping or "nuisance trapping" is then the only other alternative.

Different countries live-trap in different ways. In North America, Hancock and Bailey traps are standard (Plate 48). Nets have been used in Norway,[1] Germany, and Mongolia,[2] and seine and hoop nets also in Massachusetts.[3] In Norway, land nets, scoop nets, and diving nets have been effective, averaging a capture effort of 1.9 hours/beaver caught and capturing 84 beavers during 22 nights.[1] Live-trapping with nets is labor-intensive and does not work well in ponds clogged with branches and logs. Large cylindrical traps made from welded wire ("Belarus beaver traps"), placed on land on beaver trails, have worked well in Russia and Mongolia.[2]

What constitutes "too many beavers"? Too many for whom? Clearly, we have in mind conflict with humans. The beavers themselves are quite resilient. Do beavers suffer when their populations grow without any human intervention? Our own studies at Allegany State Park showed that body weights, family sizes, and other measures of a saturated population that had not been trapped for 20 years did not differ from other, trapped populations. Beavers adjust their reproduction to population density: in saturated populations most 2-year-olds do not breed, while in trapped populations 50% of 2-year-olds reproduce.[4]

In less developed areas, beavers are managed for fur trapping. In Newfoundland, a beaver population was maintained at a stable level when 8%–29% of the beavers were harvested annually. If less than 20%–25% of the beavers were removed per year, the trappers would suffer an economic loss when pelt prices were high.[5]

Where Should We Permit Beavers to Settle and Stay?

In short, beavers should be permitted to settle where they cause the least amount of damage. By definition, such sites are located at remote streams and lakes, far away from developed areas. On the other hand, beavers and their ponds are popular with children, ornithologists, anglers, and others. For maximal recreational and educational benefit, we welcome beavers near population centers, often in state and national parks. Careful proactive beaver management ensures a workable compromise between these two goals.

Table 21.1 | Components of Proactive Beaver Management

Build on higher ground so that flooding poses no danger.
Build road bridges with large arches rather than narrow culverts.
Where roads parallel streams, place houses on the side of the road, rather than the stream, so that stream crossings are not necessary.
Install wire devices that prevent damming up of water, such as "Beaver Deceiver"[1] or "Beaver Stopper" at culverts, or "levelers" in ponds.
Plant trees that beavers find unpalatable.
Protect valuable specimen trees with wire or paint with sand-containing paste (Plate 49).
Provide alternatives (food, wetlands) where beavers will do no harm.

Proactive Beaver Management

People living in earthquake- or avalanche-prone areas have learned to prepare themselves for catastrophic events. They build and live accordingly. Now that beavers are here to stay with us, residents near streams or wetlands should do likewise. Some of the precautions are listed in Table 21.1.

Three devices have been developed to prevent beavers from causing damage by flooding and from moving to certain areas: the Beaver Deceiver, the Beaver Stopper, and the levelers. Their design and dimensions have to be adapted to the conditions of a particular site. As we gather experience, these management tools are constantly being improved.

Beaver Deceiver

Beavers respond to the noise of rushing waters by damming up the water flow. Road culverts are prime targets for the beavers' activity. The Beaver Deceiver was developed by Skip Lisle in Maine to prevent beavers from plugging up culverts, which in turn leads to the flooding of roads. This device (Plate 50) is a fencelike exclusion cage consisting of posts and welded wire mesh. It is placed before the upstream opening of a road culvert. The Beaver Deceiver keeps beavers far away from culverts and thus prevents them from piling up branches against and into the culvert.

Not being able to reach the culvert itself, beavers at most can pile material against the wire mesh fence. This will not substantially raise the water level, as the periphery of the device is rather large, the mesh size is large, and it cannot easily be plastered with mud. The size and shape of the device vary with the local stream topography. It has to be adapted to the specific width and the curvature of the stream above the culvert. The entire length of such a fence ranges from 12 to 40 m. Generally, a trapezoidal design with three 5-m-long sides and a short one at the culvert works best.[6,7]

Figure 21.1 | Beaver Stopper.

Beaver Stopper

The Beaver Stopper is a wire mesh cylinder inserted into a culvert. To prevent beavers from plugging up this "tube," a second, larger wire mesh cylinder covers the inner one (Fig. 21.1).

Leveler

Levelers are pipes that run through a beaver dam to drain the pond in such a way that the water level cannot rise above a certain, dangerous level. Many years of experience have led to effective devices that beavers cannot plug up. The Clemson-type leveler is a better-known version (Fig. 21.2). A polyvinyl chloride (PVC) pipe is placed horizontally through a beaver dam. Its upper end extends several meters into the pond and is surrounded by wire mesh. The pipe has holes 2–5 cm in diameter so that small amounts of water enter it at many places. This reduces the noise of running water that would trigger the beavers' building activity.

Beaver Behavior We Have to Know for Effective Management

1. General behavior. Beavers are flexible in their food habits and also accept a variety of habitats, ranging from remote wilderness to shopping centers and golf courses. They adapt easily to new situations and are resilient in the face of setbacks, such as natural catastrophes and managers' actions. If we destroy lodges or dams, they generally will rebuild these structures.

Figure 21.2 | Leveler: a pipe system to keep the water in a pond below a certain level. Note the holes on the underside of the pipe. They prevent beavers from plugging up the pipe.

2. Signs that beavers are present. Some beaver signs are obvious even to the uninitiated: a felled tree, a fresh stump, a newly built dam. Others often go unnoticed: a few thin peeled sticks, or scent mounds. These telltale signs can indicate that beavers are immigrating, and to a certain extent, their general population strength. With regard to the latter, short of a detailed population census, we can glean certain diagnostic information from examining only a part of the population. The more beaver colonies there are along a stream, the more scent mounds tend to exist at any one site (see chapter 6). In "thick" populations, beavers also show more wounds and scars, presumably as signs of strife between colonies. The presence of 2- to 3-year-olds is a further sign of high population density. These are offspring that stay in the parental colony one year longer than usual because new places are hard to find. Because these animals are indistinguishable from adults, trappers have labeled these 2- to 3-year-olds "additional adults" beyond the resident breeding pair. We can identify these "supernumeraries" as 2- to 3-year-old offspring of the local pair only where beavers have been tagged individually. To check for scars and to be sure about the numbers of beavers in a colony, live-trapping is necessary.

3. Use of space. It is important to know beavers' foraging range (i.e., how far from an occupied lodge we can expect beavers to affect vegetation) and also that it is very variable, depending on the tree species available. Conversely, if

we find signs of feeding, we have to be able to estimate where these animals live. All the places visited in the course of 1 year, primarily for feeding, constitute a family's home range. The home range of younger, dispersing, or transient beavers is larger than that of established territorial familes.[8]

Managers also need to know the expected dispersal distance of young beavers, discussed in chapter 12, to anticipate where in the vicinity of existing colonies new settlers may show up.

When beavers are moved to new, more acceptable places, they are very likely to wander away from the release site and might cover long distances, up to hundreds of kilometers. To increase the chance that they will stay, it is better to transfer whole family groups and to do so during the fall season. We should also keep in mind that others will immigrate to a site that has just been cleared of its resident beavers if it remains attractive.

4. What we need to know when we trap beavers. Of course, the activity rhythm of the local beavers matters when it comes to trapping. Furthermore, the best baits are food such as aspen (*Populus* sp.) and scent such as castoreum. Different age and sex classes can exhibit different levels of bait and trap shyness. Juveniles enter traps more readily than adults, while adults respond more to scents such as castoreum. There is a tendency for males to be more easily caught than females.[9]

   The property owner or his or her experienced agent should do the trapping. Hired commercial trappers are more likely to leave some nuclear breeding population in order to be able to come back later and trap some more beavers.

5. New Approaches. In newly emerging management techniques, we try to manipulate the behavior of beavers. Redirecting their feeding behavior is the easiest intervention. To keep beavers out of areas to be protected, we can plant less palatable food and give them a choice of acceptable food somewhere nearby. To manage vegetation this way, it is necessary for several landowners to cooperate. To prevent beavers from cutting down trees, a mixture of paint and sand, pasted on the trunk, has been successful. Other chemical repellents "in the pipeline" are extracts from unpalatable plants or their active compounds, and predator odors (see chapter 14). Artificial scent marks may deter colonization in the short term, but they work only for a short time because they need reinforcement by real beavers, and they work only when the population density is low.

Fertility Control. Sterilization of beavers has been tried but has not become widespread as a management technique.[3] In Massachusetts, Brooks and coworkers[3] sterilized 1 adult in each of 10 colonies (5 males, 5 females) by tubal ligation. In addition, they castrated 1 adult in each of 4 colonies (2 males, 2 females). For comparison, 1 adult in each of 4 colonies (2 males, 2 females) was sham-operated.

Only 2 of the 14 treated colonies reproduced, while all 4 colonies with sham-operated adults produced young. Males and females sterilized by tubal ligation did not change their behavior. However, castration did have effects: One castrated female left her colony 3 weeks after treatment. In this colony a 2-year-old female immigrated and mated with the resident male. Otherwise, the behavior of castrated females was normal. But the behavior of the 2 colonies with castrated males did change. One castrated male resembled females in how he maintained his dam and lodge. The other colony relocated, and the castrated male appeared to live separately.[3] While this method may be used at single sites, it is too costly and labor-intensive to be used on a large scale.[9]

Management of Gene Pools. In both North America and Eurasia, repeated transfers and reintroductions have resulted in different subspecies becoming genetically mixed. Moreover, the North American beaver has been introduced in Europe. Wildlife managers have removed it from the Danube population in Austria, but large numbers still live in Finland, northwestern Russia, and Kamchatka in Asia.[10] In each case where genetics of a population is of concern, one has to determine the genetic composition of the population, agree on a goal, and pursue a feasible way to manage the gene pool, as has been done in Austria.

REFERENCES

1. Rosell, F., and B. Hovde. 2001. Methods of aquatic and terrestrial netting to capture Eurasian beavers. Wildlife Society Bulletin 29: 269–274.
2. Uhlenhaut, K., M. Stubbe, R. Piechocki, and N. Dawaa. 1977. Der Lebendfang des Flussbibers *Castor fiber* L. 1758. Archiv für Naturschutz und Landschaftsforschung, Berlin 3: 211–222.
3. Brooks, R. P., M. W. Fleming, and J. J. Kenelly. 1980. Beaver colony response to fertility control: evaluating a concept. Journal of Wildlife Management 44: 568–575.
4. Müller-Schwarze, D., and B. A. Schulte. 1999. Behavioral and ecological characteristics of a "climax population" of beaver (*Castor canadensis*). In: P. E. Busher and R. M. Dzięciołowski, editors. Beaver protection, management, and utilization in Europe and North America. New York: Kluwer Academic/Plenum. p 161–177.
5. Payne, N. P. 1984. Mortality rates of beaver in Newfoundland. Journal of Wildlife Management 48: 117–126.
6. Lisle, S. 1999. Resolving the human-beaver conflict, or they're here to stay. Presentation at conference titled "Beaver and Common-Sense Conflict Solutions," Estes Park, Colorado, September 8, 1999.
7. Lisle, S. 1999. Wildlife programs at the Penobscot Nation. In: Transactions of the 64th North American Wildlife and Natural Resources Conference. Washington, D.C.: Wildlife Management Institute. p 466–477.

8. Wheatley, M. 1997. Beaver, *Castor canadensis*, home range size and patterns of use in the taiga of southeastern Manitoba: III. Sex, age and family status. Canadian Field-Naturalist 111: 211–216.
9. Schulte, B. A., and D. Müller-Schwarze. 1999. Understanding North American beaver behavior as an aid to management. In: P. E. Busher and R. M. Dzięciołowski, editors. Beaver protection, management, and utilization in Europe and North America. New York: Kluwer Academic/Plenum. p 109–128.
10. Saveljev, A. P., and V. G. Safonov. 1999. The beaver in Russia and adjoining countries: recent trends in resource changes and management problems. In: P. E. Busher and R. M. Dzięciołowski, editors. Beaver protection, management, and utilization in Europe and North America. New York: Kluwer Academic/Plenum. p 17–24.

Index